# Earth Rovers, for Exploration and Environmental Monitoring

Patrick H. Stakem

© 2014, 2022

Number 1 in Robots series

# Table of Contents

Introduction..................................................................................5
    The author............................................................................12
    Dedication.............................................................................12
Robotic Explorers, off-Earth experience..............................14
    Lunar Rovers........................................................................15
        Soviet Luna missions......................................................15
        U. S. Surveyor Missions.................................................16
        Chinese Yutu Mission.....................................................17
        Chandrayaan Indian Mission...........................................17
    Google X-Prize.....................................................................17
    Lunabotics Mining Challenge..............................................18
Earth Explorers.........................................................................21
    Atmospheric satellites .........................................................21
    Volcano Explorer .................................................................21
    Archaeology applications....................................................23
    Robot Explorers in Nuclear Power Plants..........................25
    NASA Rovers in Antarctica................................................30
    Greenland Ice Pack Rover (Grover) ...................................31
    Robots Underground............................................................36
    Underground Unassisted .....................................................39
    Underground exploration sensor suite.................................45
    An approach to exploring lava tubes with CubeRovers............45
    Methods of Underground Navigation .................................46
    Underground Communications............................................48
    Weather observing robots....................................................48
    Plastic Pollution...................................................................49
    Lorax....................................................................................50
    Robotic drones in Agriculture.............................................51
    UAV's in forestry................................................................52
Domains......................................................................................52
    Ground..................................................................................53
    Underground........................................................................53

| | |
|---|---|
| Air | 54 |
| Water/underwater | 55 |
| Multi-Domain Systems | 56 |
| Cooperative systems | 57 |
|     Robot Swarms | 59 |
| Smart Sensors, and Sensornets | 60 |
| Monitoring and cross monitoring | 61 |
| Mobile platforms | 63 |
|     Advanced battery and motor technology | 64 |
|     Internet, and IoT | 65 |
|     Mobility platforms | 66 |
|     Flight Platforms | 68 |
|     Advanced battery and motor technology | 68 |
|     Embedded processors | 69 |
|         Arduino | 70 |
|         The Raspberry Pi | 72 |
|         Embedded pc  x86 architecture | 73 |
|     Software | 76 |
|     Impacts of software | 77 |
|         Open Source versus Proprietary | 78 |
|         Languages | 81 |
|         Operating system | 83 |
|     File Systems | 86 |
|         RTOS | 87 |
|     An Architectural Model | 92 |
|         NASREM | 92 |
|         Real Time Control System (RCS) | 95 |
| Standards | 97 |
| Systems Engineering Design Process | 100 |
|     Requirements | 101 |
|     Specifications | 103 |
|     Design Decomposition | 104 |
|     System Architecture | 107 |
|         Fault Tolerant Design | 111 |
|         Redundancy | 111 |
|     Safety | 112 |

- Security .................................................................... 114
- Wrap-up ......................................................................... 118
  - Getting involved; getting students involved ........................ 119
    - STEM .................................................................. 119
    - FIRST ................................................................. 120
    - NASA/GSFC Summer Engineering Robotics Boot Camp 120
    - Zero Robotics Competition ......................................... 121
    - On your own ........................................................... 122
    - Some Suggested Projects ............................................ 125
- Glossary ........................................................................ 127
- Bibliography ................................................................... 132
- Resources: ..................................................................... 143
- If you enjoyed this book, you might also be interested in some of these ........................................................................... 146

# Introduction

This book introduces and formalizes the topic of using increasingly capable autonomous robotic explorers on our home planet. We have long used robotic spacecraft to explore space, our moon, the other planets of our solar system, and their moons. At least one of our spacecraft, Voyager, has left the solar system, and is still returning data on interstellar space. Unmanned explorers precede human ventures into space, and have been to many more distant locations than we have.

Our own planet has been extensively but not exhaustively explored by humans, from the depths of the sea to the highest layers of the atmosphere, and to the "four corners" of the Earth. Now, with the emergence and maturation of advanced robotic technology and "big data" techniques, we can do more. It is increasingly valuable to collect, store, and manage massive databases of information on our own planet. These data can be used to develop and refine models of our environment and ecosystem, which may be critical to our survival.

Is our planet heating up or cooling down? How thick are the polar ice sheets, and are they growing or receding? How much greenhouse gas is generated by human endeavors, and can the rate be controlled? When is the next volcanic eruption? What don't we know about our home planet?

Since 1957, we have used unmanned spacecraft to explore the Earth. From their unique vantage point, a global view is available. There are now large numbers of Earth-viewing scientific spacecraft focusing on different features of the environment. At the same time, explorations of the deep oceans have used remotely operated vehicles, and our understanding of the ocean surfaces have been enhanced by flotillas of autonomous buoys. Ground-based exploration and environmental monitoring have relied on human-based systems, although virtual exploration, using large compiled databases that can be mined for data also enhance the effectiveness of the ground observations. In certain environments, only an autonomous system can provide data gathering and exploration. This can be seen in the activities of the Mars Explorer Rovers. The case on Earth is much simpler. The communications delays are minimized, and the vehicle can generally be retrieved for repair and refurbishment.

We can use the accumulated the accumulated knowledge of robotic exploration (mostly at NASA-JPL) to build similar units for checking out our home. One applicable unit is the Mars Helicopter Scout, whicvh went to Mars with the Perseverance Rover. As a technology demo, it explored Jezero Crater, ahead of the rover.

# Earth based Rovers

As one example, an Earth-based Rover system was designed for exploration of the Greenland Ice Shelf. The depth of the ice is of interest, and historical trends indicate whether the ice depth is increasing or decreasing. The measurement is done with ice-penetrating radar. The tracked vehicle is both solar and wind-powered, and can operate autonomously for several weeks. Large data sets are stored in onboard storage, and the vehicle maintains bidirectional low bandwidth burst communication using satellite systems. The GPS infrastructure coupled with inertial navigation allows for path planning. I was fortunate to work on this Project at NASA/GSFC for two summers, heading teams of enthusiastic and talented college studies from around the world. There are further details later in this book. Another approach to path planning involves and eye-in-the-sky, a drone helping to plot a way around obstacles.

Such an instrument platform can be considered a satellite with a zero perigee. Designed by spacecraft engineers, the Greenland Rover (Grover) has a bus, the mobility system, and a science payload.

Another application of great interest is the Antarctic Meteor Finder (AMF). Some serendipitous samples of meteors from Antarctica have proven to be from Mars. Some recent findings of life-precursor chemicals may point to the possibility of life having once existing on the Red planet. These meteors are particularly easy to find - they are the small black objects on the large white field. The challenge is the extent of the ice field. Small objects cannot necessarily be seen from even drone aircraft. The AMF wanders the ice field, working 24x7, gathering small samples, and marking the GPS location of larger samples for later retrieval. Similar meteor-rich regions exist in the Sahara.

Earth Rovers are not limited to a positive perigee. Underwater rovers can be used to explore the oceans and the ice-water interface, where a glacier or ice sheet meets the sea floor. Very little is known about the dynamics at this interface.

In the Arctic, a tel-presence ROV named Nereid Under-Ice, and was developed at Woods Hole Oceanographic Institution. It includes high definition video, a manipulator arm, and a large sensor suite. It completed its first expedition in 2014. It travels up to 40 km from the mother-ship, and is rated to a depth of 2,000 meters. It uses a lithium-ion battery pack It does both acoustic and optical imaging. All of the sensed data is stored on board, and delivered in real time to the surface. A Fiber optic tether for communications is used, implementing gigabyte Ethernet. There is the distinct possibility that remote operation of the ROV from anywhere on the planet can be implemented, so guest scientists can remain at their home institutions (cozy and warm).

Reference: http://www.whoi.edu/main/nereid-under-ice

The U.S. Navy is already a big user of underwater drones for sampling, mine and traffic detection, using units such as the Fathom One and Sand Shark, from General Dynamics. These can act as smart buoys on the surface, or as deep submersible systems.

Carnegie Mellon's robot research group has shown the feasibility of having a robot descend into an active volcano.

Teams of non-homogeneous co-operating robots can optimize the search for data or items of interest. A ground-based rover can be vectored to a position of interest by an autonomous air vehicle. Swarms of small robots cover larger areas, and come together as required for items of interest. Smaller vehicles can be deployed from larger platforms as well. This might include dropping a ground vehicle from an air vehicle, or deploying a smaller land vehicle from a larger platform.

The technologies developed for planetary exploration can be applied at Earth. Extreme environments on Earth are not well understood or documented. On Earth, we have the advantage of infrastructure that can support exploration systems, such as GPS for positioning, and satellite-based communication with low delays. Not to mention we have an atmosphere we can fly in.

Robots can operate autonomously, on their own, or via teleoperation, with a person in the loop. With communication delays of more than several seconds or so, teleoperation becomes difficult for the human operator. The best autonomous robots are operating on

Mars. They receive top-level goals from Earth, and carry out those goals on their own. This requires a much greater sophistication.

According to JPL, when the rovers are navigating, they get a command telling them where to end up, and then evaluate the local terrain with stereo imaging to choose the best way to get there. They must avoid any obstacles they identify. As of mid-August, 2004, Opportunity had used auto-navigation to drive for 230 meters and Spirit for over 1250 meters, as part of the 3000-meter drive to the Columbia Hills.

The auto-navigation system takes images of the nearby terrain using one of Rover stereo camera pairs. These are body-mounted on Spirit, and mast-mounted on Opportunity. After stereo images are taken, 3-D terrain maps are generated automatically by the rover software. Traversability and safety is then determined from the height and density of rocks or steps, excessive tilts and roughness of the terrain. Typically, dozens of possible paths are considered before the rover chooses the shortest, safest path toward the programmed geographical goal. The rover then drives between 0.5 and 2 meters (1.6 and 6.6 feet) closer to its goal, depending on how many obstacles are nearby. The whole process repeats until it either reaches its goal, or is commanded to stop.

The Exploration Rover's autonomous driving software is more advanced than Sojourner's in several ways. Sojourner's onboard safety system also looked for obstacles, but could only measure 20 points at each step. Spirit and Opportunity typically measure more than 16,000 points from each pair of images. The average Mars Exploration Rover obstacle-avoidance driving speed of nearly 34 meters (about 112 feet) per hour is ten times faster than Sojourner's. During its entire three-month mission, Sojourner drove just a little more than 100 meters (328 feet) total. Spirit and Opportunity each broke that record in a single day; Spirit drove 124 meters during sol 125, and Opportunity 141 meters during sol 82.

The Mars Exploration Rover's Visual Odometry software system is another improvement As the rovers drive over sandy and rocky terrains, they can unpredictably slip The Visual Odometry system helps by giving the rover a much better notion of how far it has actually traveled. It works by comparing pictures taken before and after a short drive, automatically finding dozens of features in the terrain (for example: rocks, rover tracks and sand dunes), and tracking their motion between images. Combining that with the 3-D terrain shape is more than enough information to let the rover figure how it really moved, much more precisely than simply counting how much its wheels have turned.

The convergence of multiple technologies, along with the rapidly decreasing cost of highly capable systems, has led to University, high school, and even individual efforts. We are at the point with CubeSats where an individual can reasonably consider having his own payload launched into Earth orbit.

We'll discuss the application of robot explorers off the surface of the Earth, and particularly on the surface of other planets and their moons, for lessons learned. Then, we'll look at existing applications for robotic explorers on Earth. We will discuss the different domains and environments, and look at multiple cooperating systems. Then, we will explore the enabling technologies that have made these robot explorers feasible technologically, but also in terms of cost. A glossary of terms, and a set of references is included.

This is an exciting time in terms of exploration of our planet. We can consider sending out our robot minions, and view the data from the comfort of our armchairs. This is safer and easier than how the majority of our working knowledge of this planet has been obtained to date.

## The author

The author has a BSEE in Electrical Engineering from Carnegie-Mellon University, and Masters Degrees in Applied Physics and Computer Science from the Johns Hopkins University. During a career as a NASA support contractor from 1971 to 2013, he worked at all of the NASA Centers. He served as a mentor for the NASA/GSFC Summer Robotics Engineering Boot Camp at GSFC for 2 years.

He developed and ta courses for Loyola University in Maryland, the Graduate Department of Computer Science; the Johns Hopkins University, Engineering for Professionals Program; and Capitol Technology University.
He can be found on Facebook, and LinkedIn.

# Dedication

To James S. Albus, PhD, for his contributions to Robotics with his work at NIST and NASA, and for his insights on what the technology implies. He showed us how to get the work done, back when it was hard to do it.

# Robotic Explorers, off-Earth experience

We will first look at some robotic exploration vehicles used off-Earth. The major differences between these and something designed for the Earth is the environment, which is unique in each instance. At the moment, the most active rover systems are on Mars. We can look at their approaches, and take them for lessons learned. There

One could actually characterize satellites in general as robotic explorers. In Earth orbit, we have numerous weather observers, as well as environment spacecraft. We have sent similar observing spacecraft to the other planets of our solar system, putting some in orbit around them, and others onto their surfaces. There's a lot more exploration to do in our solar system.

Technologies for autonomous mobility enable the rovers to make decisions and avoid hazards on their own.

Robotic Explorers can cover larger surface territories without being subject to Orbital Mechanics, and can identify and visit targets of interest. They can hunker down in one location for a while, recharging their batteries, and take data for an extended period. They can touch the samples, pick them up, blast them with a laser, and examine the spectrum of the vapor. On other planets, the tricky part is usually surviving the landing.

In many cases, we know more about the surface of Mars than, say, the depths of the Earth's ocean, or even a volcanic vent. There's still a lot of Earth to explore yet. The easy parts are done.

## Lunar Rovers

Early in the era of space exploration, a series of rover vehicles were sent to the Earth's moon. These were designed as precursors to a manned visit. From the mid-1960's through 1976, there were some 65 unmanned landings on the moon. Now, this is the subject of a private effort, the Google X-prize. The moon is still the subject of intense study, with missions from the United States, Russia, China, India, the European Union, and Japan. In this section, we focus on rovers of the Lunar surface.

Soviet Luna missions

The Soviet Union launched a series of successful lunar landers, sample return missions, and lunar rovers. The Lunokhod missions, from 1969 through 1977, put a series of remotely controlled vehicles on the lunar surface. Lunokhod-1 was an 8-wheeled rover, operated from Earth. It was the first Rover to land on a body other than Earth. It deployed from the landing platform via a ramp. It was operational for 11 months. The follow-on Lunokhod-2 Rover could transmit live video from the surface, and had a series of soil property instruments. Its tracks were seen by the Lunar Reconnaissance Orbiter in 2010. The Lunokhod-3 rover was built but never launched. It resides at a museum. The first and second rovers remain on the moon, although the second rover was sold in 1993 at a Southby's auction. The buyer was Richard Garriott, son of Astronaut Owen Garriott. As of this writing, he has not picked up his property.

The initial purpose for the Lunokhod series was to scout sites for manned landings, and to serve as beacons. The rover could be used to move one Cosmonaut at a time on the surface as well. Lunokhod had a group of four television cameras, and mechanical mechanisms to test the lunar soil. There was also an X-ray fluorescence spectrometer, and a cosmic ray detector. The second unit conducted laser ranging experiments from Earth via a corner reflector, and measured local magnetic fields. The rover was driven by a team on Earth in teleoperation mode.

## U. S. Surveyor Missions

The NASA Surveyor missions of 1966-68 landed seven spacecraft on the surface of the moon, as preparation for the Apollo manned missions. Five of these were soft landings, as intended. All of these were fixed instrument platforms. Interestingly, Apollo-12 astronauts landed near Surveyor 3, and returned with some pieces. Not just souvenirs, these were used to evaluate the long term exposure of materials on the lunar surface.

Reference:
http://nssdc.gsfc.nasa.gov/planetary/lunar/surveyor.html

Chinese Yutu Mission

Yutu is the name of the Chinese Lunar Rover, and means Jade Rabbit. It was launched in December of 2013. It landed successfully on the moon, but became stationary after the second lunar night. It is a 300 pound vehicle with a selection of science instruments, including an infrared spectrometer, 4 mast-mounted cameras including a video camera, and an alpha particle x-ray spectrometer. The rover is equipped with an arm. It also carries a ground penetrating radar. It is designed to enter hibernation mode during the 2-week lunar night. It does post status updates to the Internet, and still serves as a stationary sensor platform.

Chandrayaan Indian Mission

The Indian/Russian Chandrayaan-2 mission is an orbiter and lander, with a current launch date in 2016. The design is unique in having been selected from student proposals. The lander will be a 6-wheeled, solar-powered rover.

# Google X-Prize

This is a lunar robotics competition, organized by the X-Prize Foundation in 2007, and is valid through 2015. It requires a team to develop, launch, and demonstrate a robot on the moon that travels at least 500 meters, and transmits back high definition video. The prize for this is $20 million. If accomplished, this would be the first vehicle to operate on the lunar surface since 1976, and the first non-governmental effort. Another goal is to capture images of Apollo hardware on the moon, verifying the presence of water ice, or surviving through the 2-week long lunar night.

This effort was originally to be funded by NASA, but that would have limited the competition to United States Teams. The X-Prize Foundation, funded by Google, has no such restrictions. There are twenty-five international teams officially working on this effort. The first team was selected in 2015, with a deadline of the end of 2016 for the remaining teams. The deadline to complete the mission is the end of 2017. Fifteen teams are still in competition.

Reference Alicia Chang (2007-09-14). "Google to Finance Moon Challenge Contest", Washington Post. http://www.googlelunarxprize.org/

# Lunabotics Mining Challenge

The Lunabotics Challenge was announced by NASA in 2010. Quoting from the announcement, "The Lunabotics Mining Competition is a university level competition designed to engage and retain students in Science, Technology, Engineering and Math (STEM). NASA will directly benefit from the competition by encouraging the development of innovative lunar excavation concepts from universities which may result in clever ideas and solutions that could be applied to an actual lunar excavation device or payload. The challenge is for students to design and build a remote controlled or autonomous excavator (lunabot) that can collect and deposit a minimum of 10 kg of lunar dirt within 15 minutes. The complexities of the challenge include the abrasive characteristics of the lunar surface, the weight and size limitations of the lunabot, and the ability to control the lunabot from a remote control center. Twenty two teams from around the nation are ready to compete at the Kennedy Space Center Astronaut Hall of Fame on May 27-28. Theses are annual events, with teams selected each year.

"The challenge will be conducted in a head-to-head format, in which the teams will be required to perform a competition attempt using the regolith sandbox and collector provided by NASA. NASA will fill the sandbox with simulated regolith, compact it and place rocks in it. Each competition attempt will occur sequentially. Between each competition attempt, the rocks will be removed, the regolith will be returned to a compacted state and the rocks will be returned to the sandbox. Consideration of prize awards will be based on each team's performance during the official competition attempt. All excavated mass deposited in the collector during the competition attempt will be weighed after completion of the competition attempt. The teams that excavate the first, second and third most lunar regolith mass over the minimum excavation requirement within the time limit will respectively win first, second and third place prizes."

reference:
http://www.nasa.gov/offices/education/centers/kennedy/technology/nasarmc.html

On Earth, Mining Companys are developing robots to work in dnagerous locations.

# Earth Explorers

Autonomous Earth explorers can be viewed as Zero altitude spacecraft. This is true for ground vehicles. We can explore up to the fringe of space with balloons. We can go quite high with unmanned aircraft, some of which are solar powered. We can go to the floor of the ocean with deep submersibles. This section will discuss some of the efforts in these domains to date.

## Atmospheric satellites

Blimp and solar powered long duration aircraft can serve as *atmosats*, operating on solar power for weeks, months, or years at altitudes of 60,00 feet or more. This is currently a research area, but applications include severe weather monitoring, forest fire detection and mapping, and data gathering from very remote locations of the globe. The data complements that from orbiting satellites. These vehicles can take advantage of the GPS infrastructure for location, and satellite communications for communications. Smaller units flown at lower altitudes as student projects use cell phones as the communications link.

## Volcano Explorer

The Robotics group at Carnegie-Mellon University is headed by the famed Red Whittaker, who lead the CMU team to win the DARPA Challenge. He is also heading the team focused on the Google Lunar X Prize. One of his many projects was a mobile platform, *Dante*, designed to enter an active volcano.

In December of 1992, Dante and his support team ventured to active Mount Erebus in Antarctica, 12,450 feet high, and about 800 miles from the pole. Erebus is important enough that manned attempts were made to enter the caldera, all unsuccessful. How much the volcano contributes to the hole in the ozone layer above Antarctica is not known. The ozone layer blocks ultraviolet light from the sun, and is critical to the continuance of life on Earth. The robot made the descent to the crater floor, some 850 feet from the top. Here it took temperature measurements, and gas samples. Erebus tends to erupt in a minor fashion several times a day. This was a NASA Project, supported by the National Science Foundation. The temperature proved to be around 1,100 degrees from the corrosive gases vented by the volcano.

Dante is a six-legged walking robot, weighing close to 1000 pounds, and connected to the support team outside the volcano by a tether to provide power and data, and possible retrieval if the robot becomes disabled.

In August of 1994, an upgraded version of the robot, Dante 2, explored the active Alaskan volcano, Mount Spurr. This is located some 80 miles west of Anchorage. The descent into the caldera was 650 feet. The robot was monitored from a control facility in Anchorage, via a satellite link, providing a live video feed.. Dante-2 was bulked up at 1,700 pounds, having been redesigned based on the earlier robot's lessons-learned. It was able to explore underneath a rock ledge, that had blocked an aerial view of a part of the crater. After successfully completing its mission, the robot walked its way out of the crater.

CMU Rovers have also been used in mine mapping. A rover called Groundhog went into an abandoned Pennsylvania coal mine and sent out live video to a conference on Mine Safety in 2002. They primary usage for the robots is seen as mapping. After initial tests, the concept of a wheeled rover was reconsidered, and an amphibious robot will be designed. This is because old mines are frequently flooded.

Reference: Byron Spice (2002-10-29). "CMU tries out new mine-mapping robot" Pittsburgh Post Gazette

# Archaeology applications

In large areas of interest, such as deserts and jungle, flying drones can be a major help in getting the big picture, locating areas of interest to explore. Imaging from orbit is good for a start, but doesn't necessarily include enough detail. Local exploration by flying drones is the next step.

One area that has challenged amateur archaeologists is the search for Genghis Khan's tomb. When the founder of the Mongol Empire died in August of 1227, his body was interned in a hidden location on the steppe's of Asia, according to the traditions and customs of his tribe. His body was reported to have been returned to Mongolia, possibly to his birthplace. Supposedly, the funeral escort killed every one with a knowledge of the location.

The interest in finding his grave site is high, and there is essentially a cult interest in him, as well as intense national pride in Mongolia. Mongolia has a lot of remote surface area to cover.

An international effort, using crowd sourcing, is termed, "The Valley of the Khan Project." Starting with armchair explorers sifting through large amounts of online satellite imagery by volunteers, a ground team then explores likely areas using a small aerial hexcopter. On-site teams use ground penetrating radar. This Project is sponsored in part by the National Geographic Society.

References:

"Palace of Genghis Khan unearthed". BBC. October 7, 2004.

http://www.nationalgeographic.com/explorers/projects/valley-khans-project/

http://exploration.nationalgeographic.com/mongolia/content/21st-century-approach-ground-surveying

Flying drones have also been used to locate and identify sites of archaeological interest in Peru since 2013. Due to performance issues at altitudes above 12,000 feet in the Andes, drone blimps have more recently been used. The collected data is used to compile 3-d maps that are then used to guide field surveys. These have found application in Mayan city sites in dense jungle in Central America as well.

Since 2013 drones have flown over at least six Peruvian archaeological sites, including the colonial Andean town Machu Llacta 4,000 metres (13,000 ft) above sea level.

# Robot Explorers in Nuclear Power Plants

There have been three major incidents in nuclear power plants, in which teleoperated robotic devices were used to aid in reconnaissance. The United States, Russia, and Japan all found need of these alternatives to humans. Unfortunately, the electronics on the robots are as susceptible to radiation as human tissue. But we can build more robots, and use out knowledge of radiation hardening to make them last longer.

A complete discussion of the physics of radiation damage to semiconductors is beyond the scope of this book. However, an overview of the subject is presented. The tolerance of semiconductor devices to radiation is in terms of their damage susceptibility. The problems fall into two broad categories, those caused by cumulative dose, and those transient events caused by very energetic particles. These are generated during a period of intense solar flare activity, for example.. The unit of absorbed dose of radiation is the *rad*, representing the absorption of 100 ergs of energy per gram of material. A kilo-rad is one thousand rads. At 10k rad, death in humans is almost instantaneous.

Absorbed radiation can cause temporary or permanent changes in the material. Usually, neutrons, being uncharged, do minimal damage, but energetic protons and electrons cause lattice or ionization damage in the material, and resultant parametric changes. For example, the leakage current can increase, or bit states can change. Certain technologies and manufacturing processes are known to produce devices that are less susceptible to damage than others.

Cumulative radiation dose causes a charge trapping in the semiconductor's insulating oxide layers, which manifests as a parametric change in the devices. Total dose effects may be a function of the dose rate, and annealing of the device may occur, especially at elevated temperatures. Annealing refers to the self-healing of radiation induced defects. This can take minutes to months, and is not applicable for lattice damage. The internal memory or registers are usually the most susceptible area of the chip. The gross indication of radiation damage is the increased power consumption of the device, with one researcher reporting a doubling of the power consumption at failure. In addition, failed devices could operate at a lower clock rate, leading to speculation that a key timing parameter was being affected in this case.

*Single event upsets* (seu's) are the response of the device to direct high energy isotropic flux, such as cosmic rays, or the secondary effects of high energy particles colliding with other matter (such as shielding). Large transient currents may result, causing changes in logic state (bit flips), unforeseen operation, device latch-up, or burnout. The transient currents can be monitored as an indicator of the onset of SEU problems. After SEU, the results on the operation of the processor are somewhat unpredictable. Mitigation of problems caused by SEU's involves self-test, memory scrubbing, and forced resets.

The *LET* (linear energy transfer) is a measure of the incoming particles' delivery of ionizing energy to the device. Latch-up refers to the inadvertent operation of a parasitic SCR (silicon control rectifier), triggered by ionizing radiation. In the area of latch-up, the chip can be made inherently hard due to use of the Epitaxial process for fabrication of the base layer (ref. 12). Even the use of an Epitaxial layer does not guarantee complete freedom from latch-up, however. The next step generally involves a silicon on insulator (SOI) or Silicon on Sapphire (SOS) approach, where the substrate is totally insulated, and latch-ups are not possible.

In some cases, shielding is effective, because even a few millimeters of aluminum can stop electrons and protons. However, with highly energetic or massive particles (such as alpha particles, helium nuclei), shielding can be counter-productive. When the atoms in the shielding are hit by an energetic particle, a cascade of lower energy, lower mass particle's results. These can cause as much or more damage than the original source particle.

Chernobyl

In 1986, there was a disaster at the Chernobyl Nuclear Plant in the Ukraine, then part of the Soviet Union. The major problem was the collapsed roof of the reactor. The first responders were mostly killed, either immediately or slowly. The remote-controlled bulldozers obtained from East Germany proved too heavy. Human workers in the area were limited to 90 seconds of total lifetime exposure, in which they could not get a lot done.

A French documentary, *Tank on the Moon*, presents a solution in which the original Lunokhod designers were called back from retirement, and used their already-radiation hardened electronics to duplicate the lunar design. These rovers, called STR-1's, were deployed to clear debris, and document the damage. They still eventually failed due to cumulative radiation damage to the electronics, but were useful for some time.

Interestingly, robots sent inside the facility later, discovered a new form of fungi, thriving in the high radiation environment. Samples were obtained, and its properties were studies. To quote the article referenced below,

"This slime, a collection of several fungi actually, was more than just surviving in a radioactive environment, it was actually using gamma radiation as a food source. Samples of these fungi grew significantly faster when exposed to gamma radiation at 500 times the normal background radiation level. The fungi appear to use melanin, a chemical found in human skin as well, in the same fashion as plants use chlorophyll. That is to say, the melanin molecule gets struck by a gamma ray and its chemistry is altered. This is an *amazing* discovery, no one had even suspected that something like this was possible."

Weird life forms on our planet makes you wonder what else is out there.

Reference:
http://www.scienceagogo.com/news/20070422222547data_trunc_sys.shtml

Three Mile Island

To quote from Wikipedia, "Three Mile Island Nuclear Generating Station (TMI) is a nuclear power plant (NPP) located on Three Mile Island in the Susquehanna River, south of Harrisburg, Pennsylvania in Londonderry Township. It has two separate units, known as TMI-1 and TMI-2. The plant is widely known for having been the site of the most significant accident in United States commercial nuclear energy, on March 28, 1979, when

TMI-2 suffered a partial meltdown. According to the US Nuclear Regulatory Commission, the accident resulted in no deaths or injuries to plant workers or members of nearby communities."

In March of the year that Red Whittaker received his PhD. D., the nuclear reactor at nearby Three Mile Island nearly experienced a meltdown. Within a budget of $1.5 million, Whittaker and his colleagues at Carnegie Mellon built robots to inspect and perform repairs in the reactor's damaged basement, and their experiences with that project resulted in the creation of the Field Robotics Center at Carnegie Mellon University. Whittaker's later teams would also develop robots to help with the aftermath of the nuclear reactor accident at Chernobyl in 1986. In 1987, Whittaker co-founded RedZone Robotics to develop and sell (or lease) robots that could operate in hazardous environments and situations too dangerous for humans."

# NASA Rovers in Antarctica

NASA sponsored the Robotics Institute at Carnegie Mellon University to build the *Nomad* Rover, a 4-wheeled, 1,600 pound autonomous explorer. It was deployed in Antarctica in the year 2000. It's job is to find meteorites. It turns out, a lot of the meteorites found in Antarctica are from Mars, based on their chemical composition. *Nomad* is doing the a similar job on Earth to what the Mars Rovers are doing on Mars.

*Nomad* is equipped with a laser rangefinder, high resolution cameras, onboard computer, satellite data link, and a gasoline-powered generator. It is looking for meteorites on the ice with a specific set of characteristics. If one is detected, it navigates to the target for a closer look. It has an arm with a camera and spectrometer, as well as a metal detector. If the rock meets the profile of being a potential meteorite, the GPS location is logged. The robot does not collect samples, but does sort through rock fields for items of interest.

A rock the size of a potato (Allan Hills 84001) found in Antarctica in 1984 definitely came from Mars, and had chemical and fossil evidence of life. It is yet to be proven whether this definitely shows the past presence of life on Mars.

A friend and past-professor in the Applied Space Sciences Department at Carnegie Mellon University has proved that meteorites from Mars accumulate in Antarctica. I won't stress you with the math.

# Greenland Ice Pack Rover (Grover)

This project is one in which the author had direct, hands-on experience. It was a project conducted over several summers at NASA/GSFC, by student interns. The concept was to have an autonomous vehicle collect data from an ice-penetrating radar units over large areas. The instrument was developed under an NSF Grant, and had been operated by two persons on skis, or by one person on a snowmobile. The Grover project was derived from a series of tracked sensor packages deployed in Antarctica, and conducted by a international team of 40 graduate and undergraduate students over several summers. The Program, sponsored by the GSFC Office of Education, was termed the Summer Robotics Boot Camp. The 40 students were selected from hundreds of applicants, and lead by a series of Mentors, chosen from NASA Civil Service and Contractor personnel.

The original Antarctic Rovers were small commercial tracked vehicles. The Grover was an extension to this, using a custom built aluminum chassis. It was powered by electric motors, supplied by lead acid batteries. Other battery technologies were too expensive. Waste heat from the computers was used to keep the batteries warm. The batteries were charged by large solar panels, but due to the low sun elevation at high latitudes, a hybrid wind turbine systems was added. Winds off the polar ice cap had essentially no obstruction before reaching lower latitudes, and blow continuously down the ice sheet.

The primary computational resource was a pc-class dual-core Intel ATOM-based motherboard. This interfaced to the instrument, formatted, and stored the data in a solid state drive. The volume of data collect was about 4-6 gigabytes per day, and the system was sized for 3 months of autonomous operation. Communication with the autonomous vehicle was via Iridium satellite modem, with text messages. This was adequate for command uplink, and status message downlink. For testing, the

Grover included wifi for a local area network mobile connection.

The Grover unit had a GPS for location, encoders on each motor to track distance traveled, obstacle avoidance sensors, a compass sensor, battery and motor current and voltage sensors, sun sensors, and distributed temperature sensors.

The main computer ran an application in the Python language, and various distributed embedded controllers based on the Arduino architecture were used. There was a watchdog computer for main computer problems, temperature, and anomalies. A reset could be commanded over the Iridium link.

The unit could be controlled directly with a wired pendant for testing, and through the wifi or Iridium link. It could also operated in autonomous mode, following predefined patterns.

During field tests of the unit at NASA's Wallops Flight Facility in 2011, a quadcopter unit was used to make a video of the vehicle's performance from a "god's eye" view. It was later realized this capability could be used operationally, with the rover and the copter operating cooperatively. The rover would provide a landing spot and battery recharging station for the copter, and the copter could scout ahead of the rover for ice fissures, and areas of potential interest. This would provide a level of cooperation between the two systems.

Performance in field tests is seen in the excerpt from the Test summary, 432011:

"The motors routinely produced enough power and torque to move the robot at max speed (1.2 mph) in all terrain conditions (wet sand, hard packed sand, and soft dry sand). The Solar Arrays are about 13% efficient. In partly cloudy conditions with Sun almost overhead, the sunlit panels produced an average 216 W (with both panels). The other side produced an average 65 W, from non-direct sunlight and from sunlight reflected off the ocean and sand. The average wind Speed was 13 mph, which is typical of what we anticipate for the Greenland Ice Summit. This produced an average of 78 W from two wind turbines There was absolutely no indication of any tipping of the vehicle even in 40 mph gusts, which we have also experienced with it previously. The vehicle is very stable and durable and it has enough torque as currently geared, to drag a 225lb load through the sand behind it with an additional 100W of power. The load for this chassis design without the Instrument was 360W in the Driving mode and 50W in Idle mode. We anticipate another 100W worst case for the Instrument operational mode.

The tests performed on the power systems prove that GROVER will have over a 30% duty cycle in nominal Greenland Summit conditions, for the fully loaded Instrument operational mode. We are now designing a lighter version of the vehicle with more efficient solar arrays and batteries, which we expect to be 200lbs lighter and will take 100W less power to operate. Our goal then is to get above 50% operational duty cycle for nominal conditions. Note that on the beach under conditions of

10mph wind and clear sunshine, we were power positive, so we have justification for confidence that we can achieve a minimum of 50% duty cycle in nominal Greenland conditions. This is akin to operating in the daytime only around here. Also, we can easily adjust the gear ratios to drive the vehicle faster if desired."

Reference:
http://www.nasa.gov/topics/earth/features/grover.html#.U5G2T3afblc

UAV's, or unmanned aerial vehicles, have applications in exploration. They have been used by the oil and gas industry to survey large areas for potential new resources. This generally involves geomagnetic sensing, looking for anomalies in the Earth's magnetic field which can be associated with underground cavities. In addition, drones with real-time video feeds can patrol long pipelines through remote areas looking for leaks. Similar uses can be found in the Electric Power Industry, for inspection of power lines in remote regions. There are manufacturers of commercial drones for commercial and scientific use. NASA operates an unarmed version of the General Atomics MQ-9 Reaper as part of its Environmental Research Aircraft and Sensor Technology (ERAST) Program. This program uses the aircraft for remote environmental sensing, monitoring of agriculture, severe storm tracking, and also serving as a telecomm relay platform.

Reference:

http://www.barnardmicrosystems.com/about/papers_&_presentations.html

# Robots Underground

There are two underground domains that robots can operate in. One is caves, natural formations. The other is mines, man-made. Mines ae dangerous. This mining company's noticed they could improve safety and yield with under ground diggers and transport.

Fully autonomous equipment is more expensive, but the operation does not necessarily need to be in the vicinity. This has resulted in the most productivity gains.

The "Mine of the Future" of the Rio Tinto Group takes this to extremes. Operators are in Perth, and the machinery is operating in Pilbara region of north west Australia, a distance of 1,280 km. Autonomous robots can also be used to inspect mines, providing continuous video. Rio Tinto is focused on world-wide mineral extraction. The company was formed in 1873, at a mine of the same name in Spain. It has headquarters in London, and Melbourne, Australia. The Rio Tinto river in Spain has bright red flowing water, due to acid mine drainage for the last 5,000 years or so. The company is the largest producer of aluminum in the world. It operates mines on six continents.

The company has a fleet of 80 autonomous trucks from Komatsu working in Australia. Also in Australia is their "Intelligent Mine," the Koodaider in Australia.

Global Mining Giant Rio Tinto spent $2.6 billion on the automated Koodaideri open cut iron ore mine in Pilbara, Western Australia. Production began in 2021, and is expected to continue for 30 years.

Fortescue Metals Group had an incident with one of their autonomous hauling vehicles, when it backed into a parked vehicle. It was speculated that the accident was due to a dropout in a wifi signal. No injury's occurred.

Robots can be useful in re-opening abandoned mines. This can be useful for mining materials that are in current demand, but were not traditionally used, such as certain rare earth materials.

Mining things like diamonds and rare Earth metals vital for semiconductor product are rarely found in an easily accessible location. Diamonds, in South Africa are found miles down, in hellish conditions. How hard can mining be? A group at Carnegie Melon University asked that question a few years ago. They were doing mapping work to define where the old coal mines were, so they could possibly be remediated. If they could be sealed, ground water won't be able to leach out toxic material into the water supply. If we know where a honeycomb of tunnels are, we know not to build a major structure above there.

The Robotics Team at Carnegie Mellon explored the calderas of active volcanos with some success. This is very dangerous. You can replace a robot. It wasn't a huge move to do underground tunnels.

Now, the focus is mining the moon (for ice), Mars (for water and methane), and the asteroids for rare metals. Humans won't be doing this. Our lessons learned from robot mining will be applied to this much harder problem, with a lot of experience in the topic.

One reason that underground robots are important is that they replace humans in dangerous environments. They can tirelessly map abandoned mines and cave systems, including lava tubes. They can also mine for precious metals and diamonds, deep underground. At some point, we will send them to mine asteroids, and ship the rare materials back.

Robots underground can be operated in several ways, from remote control and teleoperation to full autonomy, A lot of commercial grade excavators can be operated by a handheld control. It is not ideal, as the operator may not have the visibility and "feel" of the machine.

Going a step further, tele-operation involves operation from a more protected location. The machinery has multiple video cameras. It can be operated 24x7 without breaks by a crew of operators. Robots can also operate in deep sea mining, and hazardous atmospheres in regular mines.

We can define two types of underground voids, caves and mines. Caves are natural features, formed by the action of rainwater, a weak acid, mostly on limestone. There are referred to as solution caves. They have underground drainage. These are also referred to as karst caves. They are natural features. Another type of underground space is a lava cave. These are formed from a lava flow from a volcano. When the lava flow hardens you have an underground void. Given enough time, flora and fauna will move in, and water action will smooth out the jagged lava. Lava caves are of special importance to explorers of the Moon and Mars. They can ideally be sealed and used as habitats and storage, as well as protection from radiation.

## Underground Unassisted

The SubT Challenge began in 2018 and consisted of two tracks: the Systems track and the Virtual track, both of which are split into three subdomains, or events – the Tunnel, Urban, and Cave Circuits. The Virtual competition focuses on developing software that can participate in simulation-based events, and the Systems competition centers on physical robots that operate in real field environments.

The range of environments require different modes of locomotion and a diverse array of robots to complete complex tasks. Team CoSTAR may use wheeled and tracked robots to cover ground faster when obstacles are few obstacles or terrain is rugged.

The Tunnel Circuit took place in August 2019 in mining tunnels under Pittsburgh, with Team CoSTAR placing second; they took first in the Urban Circuit, held in February 2020 at an unfinished power plant in Elma, Washington. The Systems Competition Cave Circuit was canceled in the fall of 2020 because of COVID-19 restrictions.

The final event was held in the 4-million-square-foot (370,000-square-meter) Louisville Mega Cavern, and featured a combination of all three subdomains that DARPA has designed – from cave systems with irregular passages and large caverns to subsurface structures with complex layouts that reach several stories high.

Team CoSTAR relied on a diverse array of robots to fulfill the mission goals. They first send in robot scouts to explore the environment, then select a subset of robots best able to collectively satisfy the overall mission goals depending on their mode of locomotion.

The robots generate a live 3D map as they locate objects that represent a disaster-response and search-and-rescue scenario, such as manikins (to simulate human survivors), cellphones, and backpacks distributed throughout a large environment.

## DARPA subterranean challenge

Led by NASA JPL, Team CoSTAR will participate in the SubT final to demonstrate multi-robot autonomy in a series of tests in extreme environments.

Eight teams featuring dozens of robots from more than 30 institutions, including NASA's Jet Propulsion Laboratory in Southern California, converged in a former Kentucky limestone mine from Sept. 21 to 24 to participate in a series of complex underground scenarios. The goal: to demonstrate cutting-edge robotic autonomy capabilities and compete for the chance to win $2 million.

A DARPA program, the event marked the final contest in the Subterranean, or SubT, Challenge, which began three years prior, attracting talent from around the world. The challenge was aimed at developing autonomous robotic solutions for first responders in underground environments where GPS and direct communications are not navailable.

The technologies developed for the SubT Challenge and extreme-environment exploration on Earth also had direct applications for space exploration. The JPL-led Team demonstrated their collection of driving, walking, and flying robots that could be used to explore extreme terrains on the surface as well as inside the caves and lava tubes on other worlds, including Earth, without human assistance.

The JPL Mobility and Robotic Systems group is comprised of approximately 150 engineers working on all aspects of robotics for space exploration and related terrestrial applications. Their primary objectives are to develop technology for in-situ robotic exploration of the solar system, and apply this technology on spaceflight missions. They focus on providing capabilities for mobility across and above planetary surfaces, and manipulation for instrument placement and sample collection. This includes system architectures; mechanical and electrical design; and software for perception, control, simulation, and operations. Their competence has been demonstrated numerous times on Mars.

The robots will also produce a live 3D map as they locate objects that represent a disaster-response and search-and-rescue scenario, such as manikins (to simulate human survivors), cellphones, and backpacks distributed throughout a large environment.

Also present will be environment-specific artifacts, such as a carbon-dioxide-emitting source that mimics a gas leak in an urban setting, or a helmet in a cave setting that would indicate a nearby human presence. The team of robots must operate autonomously, for the most part, with no or limited radio contact with a single human supervisor, and the mission must be completed in one hour. The more objects they can traverse to, reach, identify, and precisely locate, the more points earned.

"It is a complex challenge for hardware and software design, but also for the diverse team that has persevered through the challenges facing us in the competition and the real world these last three years," said Benjamin Morrell, robotics technologist at JPL and perception lead on Team CoSTAR. "It's amazing to see what the team has produced, and I'm thrilled to see our system be put to the test against some of the best roboticists in the world.

I'm also excited to see how SubT will springboard further advances in enhanced autonomous robots."

DARPA subterranean challenge

Led by NASA JPL, Team CoSTAR will participate in the SubT final to demonstrate multi-robot autonomy in a series of tests in extreme environments.

Eight teams featuring dozens of robots from more than 30 institutions, including NASA's Jet Propulsion Laboratory in Southern California, converged in a former Kentucky limestone mine from Sept. 21 to 24 to participate in a series of complex underground scenarios. The goal: to demonstrate cutting-edge robotic autonomy capabilities and compete for the chance to win $2 million.

A DARPA program, the event marked the final contest in the Subterranean, or SubT, Challenge, which began three years prior, attracting talent from around the world. The challenge was aimed at developing autonomous robotic solutions for first responders in underground environments where GPS and direct communications are not navailable.

The technologies developed for the SubT Challenge and extreme-environment exploration on Earth also had direct applications for space exploration. The JPL-led Team demonstrated their collection of driving, walking, and flying robots that could be used to explore extreme terrains on the surface as well as inside the caves and lava tubes on other worlds, including Earth, without human assistance.

The robots will also produce a live 3D map as they locate objects that represent a disaster-response and search-and-rescue scenario, such as manikins (to simulate human survivors), cellphones, and backpacks distributed throughout a large environment.

Also present will be environment-specific artifacts, such as a carbon-dioxide-emitting source that mimics a gas leak in an urban setting, or a helmet in a cave setting that would indicate a nearby human presence. The team of robots must operate autonomously, for the most part, with no or limited radio contact with a single human supervisor, and the mission must be completed in one hour. The more objects they can traverse to, reach, identify, and precisely locate, the more points earned.

"It is a complex challenge for hardware and software design, but also for the diverse team that has persevered through the challenges facing us in the competition and the real world these last three years," said Benjamin Morrell, robotics technologist at JPL and perception lead on Team CoSTAR. "It's amazing to see what the team has produced, and I'm thrilled to see our system be put to the test against some of the best roboticists in the world. I'm also excited to see how SubT will springboard further advances in enhanced autonomous robots."

# Underground exploration sensor suite

Such parameters as atmospheric pressure (if any), temperature, light level, radiation detection, and such provide guidelines for a rover sensor suite. In addition, an odometer function provides distances traveled. For further details, we might deploy a ground penetrating radar on a rotation mount, so it can measure distance to the surface as well. GPR works well in dry materials and can detect changes in density up to 30 meters. Testing for gases would be a good idea, as would having the ability to collect samples of the wall and floor material. A lidar unit can provide a 3-D map of the tube, but this puts additional strain on onboard power, computation, and data storage resources.

A stereo camera with changeable focus, along with the a video processing pipeline in the computer, would be very useful. This is a standard feature in the ARM architecture chips.

# An approach to exploring lava tubes with CubeRovers.

This section discusses an approach that evolved at a summer session of undergraduate and graduate students that the author mentored. It was a Cubesat-based program, so naturally the lava tube explorer was Cubesat-based. This is not a problem, as the Cubesat specification merely presents a standard architecture, just for Earth orbit. We can use them less restrictively at the Moon. For example, you can use one in lunar orbit for a communications relay. Another Cubesat architecture can provide a lander vehicle/operations base. Individual explorers are essentially Cubesats with a mobility platform. The platform can be ground-based, with wheels, tracks, or legs.

The project defined a fixed base on the lunar surface, with a mothership/carrier that served as a recharging station and data assembler and forwarder. This unit communicated with the orbiting communications Cubesat relay. The approach of using the Cubesat architecture for all the elements allows for an economy of scale, and an overall reduction in complexity.

It also provides a commonality in architecture, and opens the possibility of implementation of a swarm-type cooperative behavior. Here, multiple co-operating units act together. This is not required for the Lunar mission, but can be useful elsewhere.

Methods of Underground Navigation

Navigation in a structured environment is not a challenge. It's the real world that's a problem. A telerobot operates in a three-dimensional world, as do we. If they are not fixed units, we need to measure a workspace's invariant's, the dimensions, axes, possible navigation and reference points. A mobile robot can use dead reckoning or a beacon-based navigation. This works well on Earth, where we have navigation grid infrastructure such as GPS, but doesn't work on Mars. And Mars has no usable magnetic field. A key consideration for navigation is the choice of a coordinate system. For hundreds of years, Earth has had a well-defined latitude/longitude grid. There will be similar systems set up for the Moon, and Mars. Navigation becomes obstacle avoidance when done up close.

Loran is a World War-II era radio frequency navigation aid, using fixed beacons. As a supplement to space-based systems such as GPS, a variation called eLoran is being deployed. This approach would be feasible on other planets as well, with solar-powered beacons.

On the surface, a rover vehicle with a camera can note the position of the Sun to get an idea where it is. Celestial navigation, using the "fixed" stars would also work. From the moon, the position of the Earth would help, and the rising and setting of Mars' two small moons would be usable.

Once a lava tube is entered, the robotic vehicle must operate autonomously, due to the lack of a communications link. We might also consider radio relays within the tube, but this gets complex fairly quickly. Tethers and trailing cables are problematic.

We can have the robot drop, say, marbles, to mark its trail. We're reminded of Daedelus getting out of the labyrinth. There is also the possibility of using spray paint to indicate which of several branches were taken.

## Underground Communications

Communications can be via a tether, or radio. A tether is handy to supply power as well, but is a nuisance due to getting snagged. Radio communications is a problem is there are numerous twists and turns in the path. Radio relays might be used, with the explorer dropping them as needed. Very low frequency radio has been shown to work through rock.

For communication with units underground, extremely low frequency systems work, if the depth is not too great. Explorer units would drop relay units from their entry point, as they go along, but there necessarily a limit to the number of these that can be included.

Since the Lava Tube explorers will, by definition, be operating underground , the opportunity for recharging is not available. The battery's have to be sized to allow sufficient operating time. We might also envision a dedicated charger unit that can recharge or recover and tow the Explorer.

Automated loading systems move excavated material into haulers. In terrestrial mining, an automated loader can load as fast as a human operator. Commercial ALS systems are available. Not related to mining, automated systems can also do the easier tasks of loading and unloading pallets.

## Weather observing robots

Flying through Hurricanes is routinely done by a dedicated cadre of pilots and crew, operating from southern Florida. The in-situ measurements they provide are invaluable to tracking and early warning of these storms. It is also incredibly dangerous. NOAA is now using unmanned instrumented drones in this role. They are expendable, have a longer working time on target, and provide more data. These units may eliminate or minimize the manned missions.

## Plastic Pollution

More and more plastic is getting into rivers, lakes, and oceans, becomiong fatal for animal life. This is becoming a big problem at Marinas, harbors, and ports around the world. The problem is only getting larger and worse. This crap can result in a " die out events" which now causes yet more pollution. The costs are high, including property damage, fines, and mass death events. One commercial outfit is Clean Earth Rovers,

Asn innovative company has come up with the Plastic Piranha, that skims the offending plastic, and returns it for recycling. It can handle about 300 pounds per trip. Think of it as a Roomba for the Ocean. Much of the Earth's oceans are filled with plastic debris.

A big problem is oxygen levels in the waters continue to drop. This is not good for marine organisms. Oxygen levels in nearshore waterways around the world are dropping like never before. This causes major fish die outs and reduced water quality.

Agricultural runoff is a direct result of pesticides, and other agricultural bi-products causing, algae blooms that kill marine life, reduce water quality, and reduce oxygen levels in the water.

# Lorax

Lorax is a project of CMU's Robotics Institute, and the name stands for Life on Ice: Robotic Antarctic Explorer. This will be an autonomous system to survey the population of microbes on the Antarctic ice sheet. Such extremophiles are known to exist on Earth. A team from Colorado University in Boulder found microbes thriving on the high slopes of Andean volcanoes, surviving heat, ultraviolet radiation, and noxious gases. Bacteria are often seen in the soil after a glacier retreats. These environments are not much different than those on Mars. This is a NASA sponsored project.

One goal of this project is autonomous operation for a month. It uses both solar power and wind power. A model was tested in 2005 on a frozen lake bed in New Hampshire. It successfully traversed 14 kilometers autonomously, and returned to its starting point. It is a 4-wheeled vehicle.

## Robotic drones in Agriculture

Every farmer would love a god's eye view of his fields. The Agribotix drone provides that data. Small drones are self-stabilizing, due to 3-axis MEMS autopilots. They can be directed to fly in a specific pattern in a defined area outlined by its GPS co-ordinates. Winged aircraft are more efficient in this role than quadcopter-type platforms.

Farmers don't necessarily want to play with aerial toys, and don't want an advanced course in remote imaging. What data does the farmer get? They want to know how their crops are doing.

Crop scouting allows the farmer to see which areas are getting enough water, and which are not, leading to optimization of irrigation, and water conservation. Field health is provided by color imagery. However, another important tool in agriculture is near-infrared imagery. This can be accomplished with available digital cameras, and filters. The visible spectrum image and the infrared image data are differenced on the ground, and displayed. This image is called the DVI, or Difference Vegetation Index. It provides a very quick indication of fertilizer efficiency. The correct amount of fertilizer can be applied to the correct areas, maximizing yield while minimizing run-off.

The imaging can also reveal the need for pesticide application, and outline the area needs. This allows controlled application, without excessive runoff into the water table.

A simple and inexpensive platform has the ability to vastly increase agricultural efficiency, while protecting the environment, in smaller farms.

## UAV's in forestry

Developed from military technology, small unmanned aerial vehicles have been used in Forest fire mapping. There are a series of these, available in the commercial market. Some are backpack-sized, to make them easier to get into rugged terrain for deployment. Most are hand launched, and use electric motors. They generally stay below 2000 feet above ground level, and can fly for an hour or more. They support real time image transmission.

# Domains

Mobile robots can operate in many domains. With difficulty, they can be designed to operate in multiple, very different domains. This section will discuss the unique challenges of the different environments.

# Ground

Ground based robot explorers can use a variety of locomotion. This includes wheels, treads, legs, etc. The Mars rovers tend to be 6 wheel vehicles, with each wheel operating independently under computer control. Legged locomotion is more complex to accomplish, but is more versatile in rough terrain. Hovercraft can be used, but they expend a lot of energy getting and staying airborne. They work well over ice, which is generally smooth, but not over sandy areas, where they create their own dust cloud. Wheeled systems are the simplest and easiest, although track is called for in difficult terrain. It is less energy efficient. Many wheeled, tracked, and legged platforms are available in the marketplace. Larger systems might be derived from wheel chairs, carts, or vehicles. Larger platforms can also use re-purposed internal-combustion powered vehicles. The larger the chassis, the more the need for built-in safety. Take this from a guy who was pinned to the wall by a 600 pound mobile platform, without the big red stop-that-now button.

# Underground

We can define two types of underground voids, caves and mines. Caves are natural features, formed by the action of rainwater, a weak acid, mostly on limestone. There are referred to as solution caves. They have underground drainage. These are also referred to as karst caves. They are natural features. Another type of underground space is a lava cave. These are formed from a lava flow from a volcano. When the lava flow hardens you have an underground void. Given enough time, flora and fauna will move in, and water action will smooth out the jagged lava. Lava caves are of special importance to explorers of the Moon and Mars. They can ideally be sealed and used as habitats and storage, as well as protection from radiation.

# Air

Flying explorers can cover a lot of ground, with an increase in the energy expended. Winged systems can glide, minimizing energy expenditure. Winged craft have found no off-planet applications so far. First, you need an atmosphere (rules out the moon), and enough atmosphere (makes Mars difficult). Balloon-borne payloads have been proposed for Jupiter, which, as far as we know, has no solid surface. For low altitude and short-to- moderate duration missions, the quad-copter or hex copter configuration is good. The energy expenditure is high, and they are not quiet.

A smaller and less expensive alternative is the Coyote drone, which has also been used in Hurricane research. Ref: http://www.livescience.com/47887-noaa-drone-tracks-hurricane.html

Lighter-than-air systems typically use helium for the lift. The are usually vertical systems, and they are at the mercy of the winds. Tethered balloon or aerostats can be also used. Very high altitude balloon platforms, free flying, can reach the fringes of space. There is sometimes a question of recovering the payload. Balloons are less expensive than heavier-than-air craft.

Aerobot is a term used for more sophisticated free-flyer packages. An Aerobot can be implemented in different technologies, heavier or lighter than air. Some lighter-than-air systems can stay aloft for 30 days or more.

## Water/underwater

Water-based systems can be limited to surface, or can be designed for sub-surface operations as well.

Smart buoys, floating data platforms, can be tethered or drifting. Vast networks of smart drifting buoys are returning data on currents, ocean temperature and salinity, and other data for that ¾ of the planet we don't have complete data for. They are generally solar powered.

I worked on smart buoys for the Coast Guard. You know what their major problem is? Not computation, not communication, not sensors – bird poop. Buoys are out there all alone on a flat watery surface, and provide a nice resting point for birds, who, unfortunately leave behind a material that hardens to concrete-like properties, and requires a grinder to remove. This is the major technological hurdle to more smart buoys.

For submersibles, the farther you go down, the more the pressure increases, and the greater the problem in keeping the water out of the vehicle. That having been said, submersibles have been to the deepest part of the ocean, and imaged weird life forms that thrive on volcanic vents. The farther you want to go down, the more expensive it is going to be. Platforms tend to be very specialized. The oil industry is a major customer for the commercial units, ROV's, or remotely operated vehicles, with grippers and arms. We would say, telerobot. The Navy has a great interest in surface underwater drones, for surveillance and ship protection. They automated vessel recognition

Interestingly, the Navy is linking the drone boats in a swarm architecture, where they can autonomously cooperate intelligently. They can share data, This was done by the Office of Naval Research, and tested for harbor defense in the Chesapeake Bay. The opaltform can use a varietry of "off-the-shelf" boats already in use. Put this on a larger platform and you have an intelligent submarine hunter.

There is an interesting open source project called OpenROV that involves a do-it-yourself fairly low cost telerobot underwater explorer. I would like someone in Australia to put a couple of these at the Great Barrier Reef, and allow teleoperation via the Internet.

Rent by the hour. They I could enjoy the reef environment without going all the way to Australia. Or, learning to swim. If one of you Aussies implement this, let me know, and give me one hour free, please.

Actually, something that is close is the *Fathom* underwater adventure Drone. It is a cloud-sourced project, and looks good.

## Multi-Domain Systems

Multiple domains systems have enhanced flexibility, at the cost of complexity. We can have, for example, a ground based system that can launch and recover a drone or aerostat. This provides a "god's eye view" of the area around the ground vehicle. This is useful for navigation and obstacle avoidance. We can also have an amphibious vehicle, that is equally mobile on the land, and water. Another multi-domain exploration vehicle might be deployed on an ice field, and drill through the ice to deploy a submersible.

## Cooperative systems

Rover systems can operate cooperatively in different domains. We are talking here of non-homogeneous units, designed to different set of requirements. We might have a ground vehicle that could deploy a smaller scout vehicle. That would be useful in underground exploration and search and rescue as well as indoors. The larger vehicle might not be able to climb stairs, for example. We might have a ground vehicle being able to launch an air vehicle, to fly ahead and define areas of interest, or of danger, for the ground vehicle. There are many advantages to having an "eye in the sky." This introduces the problem of having the air unit find the ground unit for a safe return landing. This is easily solved with a beacon system or GPS navigation. The ground system could recharge the air asset's batteries upon landing via an umbilical or induction. The air system would be short range, and would need to be capable of vertical take-off and landing; i.e., a quadcopter. The ground based systems might also deploy and retrieve a tethered balloon.

A similar approach was suggested for a Project I worked on at GSFC, GROVER, the Greenland Rover. The Project was developed and built by students, then deployed to Greenland to measure the depth of the ice sheet. The prototype can be seen at the GSFC visitor's center. We got the concept of a fly-ahead drone to looks out for crevasses too late to include it on the deployment.
In a DARPA competition, one team included a Rollocopter. It mostly rolled along the surface, but could fly over obstacles.

The concept is termed cooperative autonomy, and extends from cooperation between and among agents and systems, to human-machine interaction (as in the case of tele-operation). Virtual sensor platforms can be enabled, where different platforms cooperate on simultaneous observations.

Similarly, a waterborne systems could also make use of an air asset. The water borne system could also deploy a submersible. In one scenario, the waterborne systems goes to a predetermined location on a body of water such as a lake or bay, and deploys a small submarine craft for bottom sampling.

Drones or balloons can serve as communication relays for ground or water-based systems.

Robot Swarms

Let's talk about the birds and the bees for a moment. These groupings of animals exhibit what is termed swarm behavior, also seen in certain species of fish. This means the group moves as if one mind. So we have a group of homogeneous individuals, operating as one big virtual organism, a kind of spontaneous collective intelligence. It has its value in deterring predators, and hunting/feeding. A behavior emerges in the organisms' interactions, and their interaction with the environment.

An agent, in the computer sense, is an autonomous software entity that senses its environment, and is goal-seeking. It can also learn.

We can use this cooperative behavior model in the control of a group of homogeneous robot systems. Swarm dynamics can be applied to coordinate multi-rover systems. First, the individual robots need to be aware of each other, and should be able to communicate. We don't need any one in charge. It is ad hoc groupings of diverse individual that come together to work on the same problem. Scalability extends to very large numbers of units.

This is a research topic, beginning to be implemented and deployed. A swarm of small spacecraft has been proposed for asteroid exploration. There are a lot of asteroids, thousands of them, between the orbits of Mars and Jupiter, perhaps the debris of a planet. Each seems to be very different. The trick is to find the gold one.
Enabling Technology

The deployment of Earth rovers is enabled by rapid advances in technology. These include enhanced computational and communication capabilities, new materials, new power sources, and the commoditization of advanced technology. Moore's law continues to enhance the capability of the technology, while simultaneously lowering the price. Building-block modules of increasing complexity can be used at the college, high school, and elementary school level. Individuals have embarked on their own satellite projects, using the CubeSat concept, but the launch costs are still high. Earth based and low altitude exploration robots are within the capability and budget for many.

# Smart Sensors, and Sensornets.

*Smart sensors* include embedded processing. The IEEE Standard 1451 covers functions, communication protocols, and formats for smart sensors. Networked and wireless sensors are also covered. Moving the processing closer to the sensor offloads this task from the main computer, freeing up resources for other tasks. Sensor fusion is also applicable. This is the merging of inputs from different sensor types to achieve a better knowledge of a situation or event.

A group of sensors working together can be organized into a network. These can be an array of similar or identical sensors, or a group of sensors using different technologies to gather a more complete perspective of the sensed item of interest. The sensor network can be wired or wireless. The detection devices monitor the local conditions and perform a small local area surveillance, collect data, and translate the acquired raw data to usable information. The network can be rigidly preplanned, or ad-hoc and self-organizing. This latter approach involves swarms of sensors, not all of which need to be the same.

Sensornets are groups of autonomous (smart) sensors, distributed over a certain space. They are connected in a node-network architecture. The system can be wired, but is usually wireless, for convenience. Sensor nets have been used, for example, to monitor forest fires, and water quality. These little sensor systems have to be inexpensive, and have low power consumption. Loss of individual nodes does not greatly impact the system. A mobile platform might be a node on a sensor-net.

## Monitoring and cross monitoring

Homeostasis refers to a system that monitors, corrects, and controls its own state. Our bodies do that with our blood pressure, temperature, blood auger level, and many other parameters.

We can have the embedded processor monitor its performance, or have two identical systems monitor each other. Each approach has problems. We can also choose to "triplicate" the hardware, and use external logic to see if results differ. The idea is, two outweigh one, because the probability of a double error is less than that of a single error.

In at least one case I know of, the backup computer erroneously thought the primary machine made a mistake, and took over control. It was wrong, and caused a system failure.

To counter the effects of "bit flips" and other effects of radiation, the memory can be designed with error detection and correction (EDAC). Generally, this means a longer, encoded word that can detect n and correct M errors. There is a trade-off with price. With EDAC memory, there is a low priority background task running on the cpu that continuously reading and writing back to memory. This process, called "memory scrubbing" will catch and correct errors.

Self-test software can be included, usually running as a background task. This might also send a "heart-beat" signal to another processor or logic. At the hardware level, particularly if we are using configurable logic, we can include built-in self test (BIST).

A special-purpose timer essential for embedded applications is the Watchdog. This is a free-running timer that generates a cpu reset unless it is reset by the software. This helps to ensure that the system doesn't lock up during certain critical time periods, and the software is meeting its deadlines. This approach has saved many a system.

If the watchdog is not reset, it generates an interrupt to reset the host. This should take the system back to a baseline state, and restart it. Hopefully, normal operations will resume. The embedded system can't rely on a human operator to notice a fault in the operations or a "hung" system, and press the reset button. Many *very remote* systems, such as those in deep water or on the surface of other planets have successfully recovered from faults with a watchdog.

The watchdog timer is implemented in hardware, and does it's jobs without direct software intervention. If the software fails to reset the timer, the system reboots. This might simply reset operations and restart, or may include diagnostics before the system is restarted. So, who watches the watchdog?

# Mobile platforms

Many standard tracked, wheeled, and legged platforms are available, as well as conventional model aircraft and rotating wing aircraft on the Quad-copter pattern. Many radio-controlled models, boats, submersibles, electric aircraft, cars, and trucks are readily available and inexpensive. These serve as the mobility platforms for integrating computational, sensor, and communication packages. These might be adequate for the job, or at least the proof-of-concept prototype. This is another area where you don't have to (literally) reinvent the wheel.

Platforms can be all-wheel drive, or just several of the wheels can be driven, with idlers at the other positions for stability. One unique approach is to have a self-balancing platform with one large ball. Self-balancing mobility platforms, starting with the Segway, have made this technology off-the-shelf.

NASA's Sojourner rover uses a rocker-bogie design allows the rover to go over obstacles or through holes that are more than a wheel diameter in size. Each wheel has cleats, providing grip for climbing in soft sand and over rocks.

There is an orange filling material between the spaces in the spiral flecture , an open-cell foam called Solimide. It is cut into crescent shapes and bonded to the wheel. It fills in the open geometry of the wheels to prevent rocks and debris from interfering with drive and steering actuators. Solimide maintains its flexibility even at very low temperatures.

Actuators such as motors can themselves have built-in embedded computers. These are referred to as smart actuators. They may incorporate a local feedback and monitoring loop. IEEE-1451 is a set of standards for interfacing smart sensors and smart actuators. The standards cover functions, communication protocols, and formats.

# Advanced battery and motor technology

Batteries have continued to improve, due to new applications in cars, small electric aircraft, boats, and cell phones.

Rechargeable batteries in new chemistries are also the outgrowth of hybrid and full electric vehicles. The energy density is very high. Technologies like lithium-polymer (LioP) have expanded the operating life of equipment before recharging is required, and allow for solar recharge. These types of batteries were used in consumer electronics by 1995.

They also have the advantage of being lightweight. They also provide a higher discharge rate (greater current) than other battery technologies. However, overcharge, over discharge, and penetration can result in explosion.

Special charging circuits are required, as well as temperature and discharge current monitoring.

## Internet, and IoT

The Internet of Things is built upon web-accessible embedded systems. More and more embedded systems are on the web. This allows to integrate cheap embedded devices with ubiquitous web services, accessible with wireless technologies. An example might be smart electric meters. Smart devices, including rovers, can access data, provide data, or access services.

To make use of this concept, we need uniquely identifiable objects such as smart sensors, smart actuators, smart platforms. What is the identity scheme? The Uniform Resource Locater (URL) approach can be adopted  We also need advanced connectivity to the Internet, which provides distance-insensitive world-wide connectivity. These are large areas of the Earth's surface where the Internet does not reach, but satellite links can be used, although this is an expensive approach. The polar regions enjoy good satellite communications due to a series of polar orbiting spacecraft.

This whole thing was just getting started as of 2014. There may now be more "things" on the Internet than people. There is a huge ecosystem of devices, talking to cloud servers, and among themselves. This reduce the reliance on people (who needs us anyway?).

Cloud servers allow access to "unlimited" datasets and resources. A new trend is cloud robotics, where a connected mobile platform can offload computational and storage resources by having a good communications link.

The connectivity is enabled by the .net framework, which is open source. This allows the embedded device to be a http client. The .net framework supports most of the embedded computational architectures, including the popular Arduino and Raspberry Pi

Free and open source software and collaborative development environments enhance the deployment process. There are standard software interfaces for communication protocols.

# Mobility platforms

Many standard tracked, wheeled, and legged platforms are available, as well as conventional model aircraft and rotating wing aircraft on the Quad-copter pattern. Many radio-controlled models, boats, submersibles, electric aircraft, cars, and trucks are readily available and inexpensive. Maybe even consider a used powered wheelchair. These serve as the mobility platforms for integrating computational, sensor, and communication packages. These might be adequate for the job, or at least the proof-of-concept prototype. This is another area where you don't have to (literally) reinvent the wheel.

Platforms can be all-wheel drive, or just several of the wheels can be driven, with idlers at the other positions for stability. One unique approach is to have a self-balancing platform with one large ball. Self-balancing mobility platforms, starting with the Segway, have made this technology off-the-shelf.

Actuators such as motors can themselves have built-in embedded computers. These are referred to as smart actuators. They may incorporate a local feedback and monitoring loop. IEEE-1451 is a set of standards for interfacing smart sensors and smart actuators. The standards cover functions, communication protocols, and formats.

What platform do we need for sub-surface exploration? The Earth-based volcano explorer used wheels. Legs are possible, but present a higher difficulty is control. NASA worked with a tetrahedral explorer as a mobility platform. A snake-like crawler? More hands-on work needs to be done in terrestrial lava caves to get a good combination of mobility, stability, and low power consumption. At the moment, a robot derived from a Mars Rover design, from NASA Ames, is exploring in Lava Caves in Oregon.

## Flight Platforms

Flight platforms come in several configurations. The lighter than air craft include balloon payload, and aerostats, or blimps. For heavier than air craft, the choices are fixed wing, and rotary wing. Big advances have been made in small rotary wing craft, leading to the 4-bladed hex copter, and 6-rotor devices. The rotors can be tilted individually for attitude control, or in combination, for vertical or horizontal flight. The advantage of winged craft is that they can glide. Kite-borne payloads can also be used, but they are mostly at the mercy of the winds. Flapping wing systesm have also been demonstrated.

## Advanced battery and motor technology

Batteries have gotten better, due to new applications in cars, small electric aircraft, boats, and cell phones.

Rechargeable batteries in new chemistries are also the outgrowth of hybrid and full electric vehicles. The energy density is very high. Technologies like lithium-polymer (LioP) have expanded the operating life of equipment before recharging is required, and allow for solar recharge. These types of batteries were used in consumer electronics by 1995.

They also have the advantage of being lightweight. They also provide a higher discharge rate (greater current) than other battery technologies. However, overcharge, over discharge, and penetration can result in explosion. Special charging circuits are required, as well as temperature and discharge current monitoring.

# Embedded processors

Very-low-cost, high-performance microprocessor-based embedded systems enable wide applications. Most of these boards, complete 32-bit computers with memory and I/O, costless than $50. Add-on boards provide GPS location finding, wifi and bluetooth connectivity, 3-axis gyros, a wide variety of sensors, and motor control, via PWM.

Advances driven by cellular phones and data systems have made available small powerful processors that rival a datacenter of a few years back. They are designed for communication, and include a variety of standard interfaces. The devices are multicore, meaning there is more than one cpu. They can include specialty cores such as floating point or digital signal processing, They have memory integrated with the cpu. They support analog as well as digital interfaces. The boards tend of be deck-of-cards size or smaller, and typically cost under $50. Some examples include the Arduinos, Maple, Raspberry Pi, and BeagleBone board.

A Microcontroller is a single chip cpu (or cpu's). memory, and I/O solution. Many different variations form a single cpu architecture (such as ARM), exist, giving the designer the flexibility to choose hardware to meet his or her requirements.

This section presents and discusses some "real-world" embedded systems, at both the chip and system-level, that can be applied to robotic platforms.

Arduino

The Arduino is a simple open-source single-board microcontroller. The hardware consists of a simple open hardware design for the Arduino board with an Atmel processor and on-board I/O support. The software support includes a standard compiler and a boot loader that runs on the board, along with numerous libraries of code.

Arduino hardware can be programmed using a language similar to C++ with some simplifications and modifications, and an IDE. The Arduino project began in Italy in 2005 to produce a device for implementing student-built design projects less expensively. By mid-2011, more than 300,000 Arduino boards had been shipped.

An Arduino board consists of an 8-bit Atmel AVR microcontroller or an 32-bit ARM. An important aspect of the Arduino is the standard way that connectors are arranged, allowing the CPU board to be connected to a variety of interchangeable add-on modules called *shields*. Shields allow for interfacing with sensors and actuators, as well as general I/O. Most boards include a 5-volt linear regulator and a 16 MHz crystal oscillator although some designs dispense with the on-board voltage regulator. An Arduino's microcontroller comes with a boot loader that simplifies uploading of programs to the on-chip flash memory.

Now the Arduino architecture has been expanded into 32 bit versions, using the ARM Cortex M0 and M3. Thirty-two bits makes it easier to do complex computations.

Boards are programmed over an RS-232 serial connection. Serial Arduino boards contain a simple inverter circuit to convert between RS-232-level and TTL-level signals. Newer Arduino boards are programmed via serial communications over USB.

The Arduino board brings out the microcontroller's I/O pins for use by external circuits.

The Arduino IDE is a cross-platform application implemented in Java. It is designed to introduce programming to newcomers unfamiliar with traditional software development. It includes a code editor with features such as syntax highlighting, parenthesis matching, automatic indentation, and is also capable of compiling and uploading programs to the board with a single click. There is generally no need to edit makefiles or run programs on the command line.

The Arduino IDE comes with a C/C++ library called "Wiring", which makes many common input/output operations much easier. It uses the gnu toolchain and AVR libraries. The Atmel development Studio can also be used. Arduino programs are written in a variant of c/c++.

The Arduino hardware reference designs are distributed under an Open Source Creative Commons Attribution Share-Alike 2.5 license and are available on the Arduino Web site. Layout and production files for some versions of the Arduino hardware are also available. The source code for the IDE and the on-board library are available and released under the GPLv2 license. The Arduino design has influenced many other similar devices.

The Raspberry Pi

The ARM processor has taken an impressive place in the embedded microcontroller world. The Roomba is based on the ARM architecture. The Raspberry Pi is a small, inexpensive, single board computer based on the ARM architecture. It is targeted to the academic market. It uses the Broadcom BCM2835 system-on-a-chip, which has a 700 MHz ARM processor, a video GPU, and currently 512 M of RAM. It uses an SD card for storage. The Raspberry Pi runs the GNU/linux and FreeBSD operating systems. It was first sold in February 2012. Sales reached ½ million units by the Fall. Due to the open source nature of the software, Raspberry Pi applications and drivers can be downloaded from various sites. It requires a single power supply, and dissipates less than 5 watts. It has USB ports, and an Ethernet controller. It does not have a real-time clock, but one can easily be added. It outputs video in HDMI resolution, and supports audio output. I/O includes 8 general purpose I/O lines, UART, I2C bus, and SPI bus. The Raspberry Pi design belongs to the Raspberry Pi Foundation in the UK, which was formed to promote the study of Computer Science. The Raspberry Pi is seen as the successor to the original BBC Microcomputer by Acorn, which resulted in the ARM processor. The unit has enough resources to host an operating system such as linux.

The Raspberry Pi 2, model B+ is a powerhouse chips, with a quad-core 32-bit ARM Cortex A7, running at 900 MHz. It also has a dedicated video processing pipeline. It has 512k of sram, and uses micro-sd cards (flash memory). It has 17 digital I/O lines, and 4 serial ports. The board is 3.3 x 2.25 inches in size.

Embedded pc x86 architecture

Embedded versions of 8- and 16-bit Intel chips evolved from the general purpose cpu's. For the 32-bit versions, the 80386 family, the embedded chip version was the 80386EX. It included a static core which could run as slowly (and thus, power efficiently) as desired, down to a full halt.

The Intel ISA found its way into embedded variants of the popular 80x86 chip set, used mostly in desktop applications. These included the 80386EX, the 80387, the Vortex 86 SOC, the AMD Geode, and the ZET 80186, an open core for FPGA's. The advantage of this approach is the widespread knowledge base of the 80x86 software architecture, although the hardware architecture was not optimized for embedded applications. Most of the chips included a standard 80x86 processor core, with additional I/O and system features to reduce chip count.

As mainstream cpu's evolved for desktop and server usage, with increased speed and addressing capability, the chips became less well suited to the embedded environment. But, as family chips were developed for the laptop and tablet market, they were generally applicable in a subset of the embedded world, mostly for cost-sensitive systems with soft real time requirements, if any.

The Intel Atom cpu is an x86 architecture optimized for low power (several watts). It was introduced in 2008, and is now available in multicore and hyper-threaded editions, with speeds beyond 2 GHz. It translates x86 instructions into internal RISC instructions on the fly, and can execute two integer instructions per clock. Because the parts are IA-32 compatible, there is a large amount of available legacy software for the part. Atom processors are available on pc-architecture motherboards of a small form factor. System-on-a-chip devices are being phased in.

Nano-ITX is a pc motherboard form factor first proposed by VIA Technologies in March 2003, and implemented in late 2005. Nano-ITX boards measure 120 × 120 mm (4.7 × 4.7 in), and are fully integrated, very low power pc motherboards targeted at smart digital entertainment devices such as PVRs, set-top boxes, media centers, car PCs, and thin devices. PC operating systems such as Windows, Gnu-Linux, and bsd Unix are supported. The following photo shows a nano-ITX system, using a motherboard from Artigo. The system has 1 gigabyte of ram, 250 gigabytes of sata hard drive, 4 usb ports, and support vga and hdmi video. The power input is 12 volts. This system, configured by the author, dual-boots Windows-XP and Ubuntu linux.

Pico-ITX is a PC motherboard form factor announced by VIA Technologies in January 2007 and demonstrated later the same year at CeBIT. The Pico-ITX form factor board is 10 × 7.2 cm (3.9 × 2.8 in), which is half the area of Nano-ITX. The processor can be a VIA C7, a VIA Eden V4, a VIA Nano with speeds up to 1.5 GHz, with 128KB L1 & L2 caches. It uses DDR2 400/533 SO-DIMM memory, with support for up to 1GB. AGP Video is supplied by VIA's UniChrome Pro II GPU with built-in MPEG-2, 4, and WMV9 decoding acceleration.

PC/104 and PC/104+ are examples of standards for ready-made computer boards intended for small, low-volume embedded and ruggedized systems, mostly x86-based. These are physically small compared to a standard PC, although still quite large compared to most simple monolithic embedded systems. They often use MSDOS, Linux, NetBSD, or a true embedded real-time operating system such as MicroC/OS-II, QNX or VxWorks.

In certain applications, where small size or power efficiency are not primary concerns, the components used may be compatible with those used in general purpose x86 personal computers. Boards such as the VIA EPIA range help to bridge the gap by being PC-compatible but highly integrated, physically smaller or have other attributes making them attractive to embedded engineers. The advantage of this approach is that low-cost commodity components may be used along with the same software development tools used for general software development. Systems built in this way are still regarded as embedded since they are integrated into larger devices and fulfill a single role.

## Software

There is a variety of off-the-shelf software solutions for the small embedded processors boards. You don't have to ask, "what language do I program that in?" The choices are c-like and Java-like. Generally, you get an Integrated Development Environment that allows you to stitch together routines from code libraries.. Sometimes, you can do this graphically. You also can use the traditional coding model, for high level languages or assembly. There are many third-party development platforms that address coding for multiple architectures.

The Integrated Development Environment (IDE) is a software tool, generally hosted on a pc, to develop, download, and test code on the target embedded system. This is a set of tools for compilation, debugging, simulation, and code version control.

Usually, a rich selection of library routines are provided as well. IDE's usually include a source code editor. Some IDE's support multiple languages. The output of the IDE will be a code "load" that can be sent to the embedded system, or put into a non-volatile memory. It will include a boot loader. An IDE, hosted on a desktop machine with a large set of resources, represents a cross-tool for embedded target code development. Web-based IDE's are emerging. These run in a standard browser.

# Impacts of software

Keep in mind, executing software consumes energy and requires time. This can be observed and measured. A key issue is the development of a program style, and the development of a programming mindset; specifically. how will I debug this? This is the Design for Testability approach. It is similar to the Design for Test approach in hardware, where test points are provided at the design level.

It is critically important to document at development time. You won't have time later in the design process. The documentation can flow from requirements to specification to implementation and test. In fact, it is possible to write the documentation before the software code. It will need to be updated later to match reality, of course.

Another good practice is to define data structures first, then the processing. We all tend to focus on the algorithm first, but clever choices of data structures will simplify the algorithm. If shortcuts are required for speed or space, be sure to document your assumptions, and your violations.

Libraries of code to address specific functions; device drivers, and other software is generally available. It is always good to check whether the software function you need has already be done. It is worth a day of research, downloading, and testing to save time. However, readily available software doesn't always fit your specific problem. It is generally poorly documented, and it may contain malware.

Purchasing software from an established vendor provides some level of trustworthiness but doesn't guarantee success. Look for software modules and libraries that are supported. Software tools are also available in proprietary and open source versions.

Open Source versus Proprietary

This is a topic we need to discuss before we get very far into software. It is not a technical topic, but concerns your right to use (and/or own, modify) software. It's those software licenses you click to agree with, and never read. That's what the intellectual property lawyers are betting on.

Software and software tools are available in proprietary and open source versions. Open source software is free and widely available, and may be incorporated into your system. It is available under license, which generally says that you can use it, but derivative products must be made available under the same license. This presents a problem if it is mixed with purchased, licensed commercial software, or a level of exclusivity is required. Major government agencies such as the Department of Defense and NASA have policies related to the use of Open Source software.

Adapting a commercial or open source operating system to a particular problem domain can be tricky. Usually, the commercial operating systems need to be used "as-is" and the source code is not available. The software can usually be configured between well-defined limits, but there will be no visibility of the internal workings. For the open source situation, there will be a multitude of source code modules and libraries that can be configured and customized, but the process is complex. The user can also write new modules in this case.

Large corporations or government agencies sometimes have problems incorporating open source products into their projects. Open Source did not fit the model of how they have done business traditionally. They are issues and lingering doubts. Many Federal agencies have developed Open Source policies. NASA has created an open source license, the NASA Open Source Agreement (NOSA), to address these issues. It has released software under this license, but the Free Software Foundation had some issues with the terms of the license. The Open Source Initiative (www.opensource.org) maintains the definition of Open Source, and certifies licenses such as the NOSA.

The GNU General Public License (GPL) is the most widely used free software license. It guarantees end users the freedoms to use, study, share, copy, and modify the software. Software that ensures that these rights are retained is called free software. The license was originally written by Richard Stallman of the Free Software Foundation (FSF) for the GNU project in 1989. The GPL is a *copyleft* license, which means that derived works can only be distributed under the same license terms. This is in distinction to permissive free software licenses, of which the BSD licenses are the standard examples. Copyleft is in counterpoint to traditional copyright. Proprietary software "poisons" free software, and cannot be included or integrated with it, without abandoned the GPL. The GPL covers the GNU/linux operating systems and most of the GNU/linux-based applications.

A Vendor's software tools and operating system or application code is usually proprietary intellectual property. It is unusual to get the source code to examine, at least without binding legal documents and additional funds. Along with this, you do get the vendor support. An alternative is open source code, which is in the public domain. There are a series of licenses covering open source code usage, including the Creative Commons License, the gnu public license, copyleft, and others. Open Source describes a collaborative environment for development and testing. Use of open source code carries with it an implied responsibility to "pay back" to the community. Open Source is not necessarily free.

The Open source philosophy is sometimes at odds with the rigidized procedures evolved to ensure software performance and reliability. Offsetting this is the increased visibility into the internals of the software packages, and control over the entire software package. Besides application code, operating systems such as GNU/linux and bsd can be open source. The programming language Python is open source. The popular web server Apache is also open source.

Languages

The *c language* is an ANSI and ISO standard. Many embedded C environments differ from pure ANSI C, and only provide subsets of the language. They also provide extensions which allow more direct control over hardware. Aspects of C which do not fit target architecture well are left out.

*Java* is an object-oriented language with a syntax similar to that of c. The language is compiled to bytecodes which are executed by a Java Virtual Machine (JVM). The JVM is hosted on the computer hardware, and is an instruction interpreter program. Thus, the Java language is independent of the hardware it executes on. The JVM has also been instantiated directly in hardware.

The *JVM* is a software environment that allows bytecodes to be executed. There are standard libraries to implement the applications programming interface (API). These implement the Java runtime environment. Other languages besides Java can be compiled into bytecode, notably Pascal, ADA, and Python. JVM is written in the c language.

The JVM can emulate and interpret the instruction set, or use a technique called *Just in Time* (JIT) compilation. The latter approach provides greater speed. The JVM also validates the bytecodes before execution.

The bytecode is interpreted or compiled. Java includes an API to make up the Java runtime environment. Oracle Corporation owns Java, but allows use of the trademark, as long as the products adhere to the JVM Specification. The JVM implements a stack-based architecture. Code executes as privileged or unprivileged, which limits access to some resources.

*Python* is a general purpose higher order language. It is open source, and designed to be highly readable. It comes with most Gnu-Linux distributions now. There are many interpreters and compilers available for Python. It can be used as an object-oriented or function/procedural language. Python has expressions similar to those of Java, and there is a large standard library of routines.

In embedded, you are working very close to the hardware. At times, you may need to delve into assembly language. You may need to write a device driver (horrors!). As opposed to general languages such as c or Java, the assembly language is unique to the hardware architecture. The concepts are generally the same across assemblers for different architectures. A statement in assembly usually maps directly to one machine language instruction, where a statement in a higher order language would result in multiple machine language instructions.

Operating system

An *operating system* (OS) is a software program that manages computer hardware and software resources, and provides common services for execution of various application programs. Without an operating system, a user cannot run an application program on their computer, unless the application program is itself self-booting.

For hardware functions such as input, output, and memory allocation, the operating system acts as an intermediary between application programs and the computer hardware, although the application code is usually executed directly by the hardware and will frequently call the OS or be interrupted by it. Operating systems are found on almost any device that contains a computer. The operating system functions need to be addressed by software (or possibly hardware), even if there is no entity that we can point to, called the Operating System. In simple, usually single-task programs, there might not be an operating system per se, but the functionality is still part of the overall software.

An operating system manages computer resources, including:
- Memory.
- I/O.

- Interrupts.
- Tasks/processes/application programs.

The operating system arbitrates and enforces priorities. If there are not multiple software entities to arbitrate among, the job is simpler. An operating system can be off-the-shelf commercial or open source code, or the application software developer can decide to build his or her own. To avoid unnecessary reinvention of the wheel an available product is usually chosen. Operating systems are usually large and complex pieces of software. This is because they have to be generic in function, as the originator does not know what application space it will be used in. Operating systems for desktop/network/server application are usually not applicable for embedded applications. Mostly they are too large, having many components that will not be needed (such as the human interface), and they do not address the real-time requirements of the embedded domain.

Adapting a commercial or open source operating system to a particular embedded domain can be tricky. Usually, the commercial operating systems need to be used "as-is" and the source code is not available. The software can usually be configured between well-defined limits, but there will be no visibility of the internal workings. For the open source situation, there will be a multitude of source code modules and libraries that can be configured and customized, but the process is complex. The user can also write new modules in this case.

Operating Systems designed for the desktop are not necessarily suited to the embedded space. There were developed under the assumption that whatever memory is required will be available, and real-time operation with hard deadlines is not required.

*Real-time operating systems*, as opposed to those addressing desktop, tablet, and server applications, emphasize predictability and consistency rather than throughput and low latencies. Determinism is probably the most important feature in a real-time operating system.

A microkernel operating system is ideally suited to embedded systems. It is slimmed down to include only those features needed, with no additional code. Barebones is the term sometimes used. The microkernel handles memory management, threads, and communication between processes. It has device drivers for only those devices present. The operating systems may have to be recompiled when new devices are added. A file system, if required, is run in user space. MINIX, as an example of a streamlined kernel, has about 6,000 lines of code.

# File Systems

A file system provides a way to organize data in a standard format. An embedded system, like a digital camera, can store and organize its data (photos) and exchange the data directly with a computer. The file system stores the data, and metadata (data about the data) such as date, time, permissions, etc. Some operating systems support multiple file systems.

The important thing about a file systems for embedded systems is, don't reinvent the wheel! There are many good file systems out there, and the provide a compatibility across platforms. Most are based on the original disk operating system (dos) model.

The legacy DOS file structure is built upon linked lists. The directory file contains lists of files and information about them. It uses a 32-byte entry per file, containing the file name, extension, attributes, date and time, and the starting location of the file on disk.

The File Allocation Table (FAT) is a built map of allocated clusters on the disk. A cluster is the default unit of storage. It's size is a trade-off between efficiency of storage, and efficiency of access. A size of 256 bytes to 1024 bytes worked well in the early days. Two copies of the FAT are kept by the system, and these are on fixed locations of the storage media.

A directory file has entries for all of the files on the disk. The name of the file is in 8.3 format, meaning an 8 character file name, and a 3-character extension. The extension tells the type of the file, executable program, word processing, etc. By DOS convention, when a file is erased, the first character of the name is changed to the character E516. The data is not lost at this point. If nothing else happens in the mean-time, the file can be un-erased, and recovered. However, the E5 signifies the space the file occupied is now available for use.

Various file attribute bits are kept. The file can be marked as read-only, hidden, reserved system type, and bits indicate a directory field, a volume label (name of a storage volume, like, "disk1"), and whether the file has been archived (saved).

The FAT was originally 12-bits, but later extended to 16. Eventually, this was extended to 32-bits for Microsoft Windows. Entries in the FAT map the clusters on the storage media. These indicate used, available, bad, and reserved clusters.

## RTOS

In a real-time system, the timing of the result is as important as the logical correctness. Embedded systems find themselves in these situations a lot. There are two types of deadlines, hard and soft, and various scheduling policies to address these. A scheduling policy should have the ability to meet all deadlines. The scheduling overhead should be minimal.

In soft real time, the average performance or response time is emphasized. Desktops and servers can meet soft real time requirements. Missing a deadline is not necessarily catastrophic. Embedded examples include an elevator controller, vending machines, gas pumps, cash registers and POS, thermostats, mobile phones, and a bike computer. Missing a deadline may result in a degradation of service, but not a failure.

In hard real time, on the other hand, critical sections of code have absolute deadlines, regardless of how busy the system is. Missing a deadline means system failure. Response times must be deterministic. Examples of hard real time systems include avionics fly-by-wire system, antilock brakes, stability control in automotive applications, and nuclear power plant safety systems.

Interestingly, meeting a deadline early may be just as bad as meeting it late. There are constraint requirements on the response time for the systems.

We can have systems with the characteristics of both; these multi-rate systems handle operations and deadlines at varying rates.

*Non-Real Time* (NRT) systems are fair; they provide resources (time, I/O) to all users or programs on an equal, or pre-determined priority basis. They can arbitrate resource allocation to maximize the number of deadlines met, or minimize lateness, or some combination. Everyone gets a turn. NRT systems have high throughput and fast average response.

Android

The *Android* operating system by Google has found application in numerous smartphone and tablet computers since its introduction in 2008. It is an Open Source product based on Gnu-Linux, although not all of the code is covered by Open Source licenses. It has evolved into versions for set-top boxes, robotics, digital cameras, and digital television applications. Android supports several hardware computing platforms including ARM, POWER, x86, and MIPS.

Like Java, Android provides a virtual machine execution engine for a specific hardware platform. This virtual machine is termed Dalvik. It's strengths are in memory-limited systems, and those with hard real time requirements. Android is targeted to user input from touch, with a screen using icons. In an embedded application, it may have no direct user interface. Android uses the Gnu-Linux kernel, plus middleware, libraries of code, and API's. The user community supports a large library of applications for Android. Android has built-in support for power management.

Real Time and embedded Linux

There are several approaches to make GNU/Linux a real-time operating system. One version developed by FSM labs, and used by VxWorks, is a hard real-time RTOS microkernel that runs the entire Gnu-Linux operating system as a fully preemptive process. To address soft real-time, the GNU/Linux kernel can be modified by several available patches to add non-preemption and low latency, with a deterministic scheduler.

The standard GNU/Linux (or BSD) kernel is not pre-emptable. This means kernel code runs to completion.

The run time is not bounded, which interferes with responding to time-critical events. It is important to keep in mind that the Gnu-Linux kernel was not designed for non-preemption, as a true real-time operating system would be. Preemption has overhead, and influences throughput, usually adversely. There is a real-time Linux Foundation (.org) that is a good source of information on these topics.

Ubuntu Mobile and Embedded are variations of the Ubuntu Linux distribution for Mobile Phones, and embedded applications in general.

LynxOS

The LynxOS RTOS is a Unix-like real-time operating system from LynuxWorks It is a real-time POSIX operating system for embedded applications. LynxOS components are designed for absolute determinism (hard real-time performance), which means that they respond within a known period of time. Predictable response times are ensured even in the presence of heavy I/O due to the kernel's unique threading model, which allows interrupt routines to be extremely short and fast. LynuxWorks has a specialized version of LynxOS called LynxOS-178, especially for use in avionics applications that require certification to industry standards such as DO-178B.

QNX

QNX is a real-time operating system based on Unix. QNX Neutrino RTOS is SMP capable, and supports POSIX APIs. It is not open source.

The QNX microkernel contains only CPU scheduling, inter-process communication, interrupt redirection, and timers. Everything else runs as a user process, including a special process known as *proc,* which performs process creation, and memory management by operating in conjunction with the microkernel. There are no device drivers in the kernel. The network stack is based on NetBSD code.

RTEMS

RTEMS is the Real-Time Executive for Multiprocessor Systems, designed for embedded use, and free and open source. It is POSIX compliant. The TCP/IP stack from FreeBSD is included. RTEMS does not provide memory management, but is single process, multithreaded. Numerous file systems are supported. RTEMS is available for the ARM, Atmel AVR, and a wide variety of other popular embedded cpu's and DSP's. An RTEMS system is currently in orbit around Mars.

FreeRTOS

FreeRTOS is an open source product, with a reported application in millions of devices. The kernel is very small, on the order of 6-12k. Being an open source product, contributions form users expand on its application. Many user submitted Board support

Packages (BSP's) are available. It includes support for a file system, TCP/IP, and Networking security.

ref: http://www.freertos.org/

# An Architectural Model

### NASREM

The NASA/NBS Standard Reference Model for Telerobot Control System Architecture was evolved as a model for the implementation of advanced control architectures.

The NBS architecture is a generic framework in which to implement intelligence of a telerobotic device. It was developed over a decade as part of a research program in industrial robotics at NBS (now. NIST) in which over $25 million was spent. The NBS program involved over fifty professionals and extensive facilities, including robots, a supercomputer, mainframes. minicomputers. microcomputers. LISP machines. and AI workstations. This model, designed originally for industrial robots. is the mechanism by which sensors. expert systems. and controls are linked and operated such that a system behaves with some measure of autonomy, if not intelligence.

Systems designed from this model perform complex real-time tasks in the presence of sensory input from a variety of sensors. They decomposes high level goals into low level actions. making real-time decisions in the presence of noise and conflicting demands on resources. The model provides a framework for linking artificial intelligence. expert system. and neural techniques with classical real-time control. Sensors are interfaced to controls through a hierarchically-structured real-time world model. The world model integrates current sensory data with a priori knowledge to provide the control system with a current best estimate of the state of the system.

NASREM is a generic hierarchical structured functional model for the overall system. The hierarchical nature makes it ideal for telerobot systems, and for gradual evolution of the system. The model also provides a set of common reference terminology, which can enable the construction of a database. It defines interfaces, which allows for modularization. The model allows for evolutionary growth, while providing a structure of the interleaving of human:robotic control.

NASREM's 6-level model operates from a global memory (or database). At each level we have three processes, sensory processing world modeling, and task decomposition (execute). At the very lowest level, we have the raw sensors and the servo systems. Going up from that, we have the primitive level, the elementary move level, the task level, the service bay level, and the mission level. At the servo level, we would find cameras, and their associated pan/tilt control as well as mobility and joint motor control, with associated position feedback. At the primitive move level, we would find the camera subsystem, the arm, the mobility subsystem, and the grippers. At the elementary (or e-) move level, we would find systems such as perception or manipulation. At the task level, we might locate the entire telerobotic system.

The world modeling process starts with a sparse database. Sensor data, appropriate to the level flows in, and there might be a capability for data fusion. A task planner task can make "what-if" queries of the world model (which is state-based). The modeling task uses a global database of state variable, lists, maps and knowledge bases to allow a modeling process to update and predict states, to evaluate current states and possible states, and to report results to a task executor task. The World model, evaluates states, both existing states as evidenced by sensor data, and possible states, as postulated by the task planner.

The timing and time horizon of the various levels of the model is are vastly different. The servo level operates on the millisecond level, the primitive level, at 10's to 100's of milliseconds, and the e-move level at about a one second update interval. It would have about a 30 second planning horizon. The task level would have update interval on the order of seconds to 10's of seconds, with a planning horizon in the 10's of seconds. Moving up, the service by level would update in the 1's of seconds, with a planning horizon the order of minutes to 10's of minutes. Finally, the mission level might update on the order of minutes, with a horizon of an hour.

The servo level would accept Cartesian trajectory points from the next level up, and transform these to drive voltages or current for the mechanisms. The Primitive level would accept pose (or collection of joint angles and positions) information from the next higher level, and generate the Cartesian trajectory point to pass down the hierarchy. These involve dynamics calculations. The e-move level would accept elementary move commands and generate pose commands, after orientations in the coordinate frame, singularities, and clearances. It uses simple if-then state transition rules. The task level, the one the telerobot would be located at, accepts task commands (from the human operator), does subsystem assignments and scheduling, and generates a series of e-moves.

Real Time Control System (RCS)

RCS evolved form NASREM over decades, starting in the 1970's It is currently at RCS Level 4. RCS is a Reference Model Architecture for real-time control. It provides a framework for implementation in terms of a hierarchical control model derived from best theory and best practices. RCS was heavily influenced by the understanding of the biological cerebellum. NIST maintains a library of RCS software listings, scripts and tools, in ADA, Java, and C++.

An abstraction, the perfect joint accepts analog or digital torque commands, and produces the required torque via a dc motor. It also provides state feedback in the form of force, torque, angle or position, (depending on whether the joint configuration is Cartesian or revolute), and possibly rate. The perfect joint includes a pulse width modulator (pwm), a motor, and possibly a gearbox. Internal feedback and compensation is provided to compensate for gearbox or other irregularities such as hysteresis or stiction, For example, the torque pulses common to harmonic drives can be compensated for within the perfect joint. The perfect joint is part of the lowest NASREM level. The processing provided theoretically achieves a "perfect" torque, where the outputted torque matches the commanded torque.

The Individual Joint Controller (IJC) implements a simple control law to allow joint by joint operation of the manipulator.

The IJC provides a functional redundancy to the higher level telerobot control. It accepts inputs from a kinematically similar mini-master controller. This simplifies the computational requirements on the IJC, by removing the need for coordinate transformations. The IJC does not include any dynamic joint coupling compensation. It basically implements seven parallel, non-interacting control laws, that may be simple PD loops.

A telerobot control system can be implemented the first 3 (of 7) levels of the NASREM model. Further levels could be added later in a phased evolution of the system. For early systems, the human operator provided the functionality of the upper control levels.

The telerobot controller initially implemented the first three NASREM levels, and could accept commands from a joystick-type element, a mini-master, or higher levels of the model. This level required a computational capability of several MIPS, and an accuracy of 32 bits. Floating point capability was assumed. This controller could perform coordinate transformations in real time, although the computation burden argued for a custom hardware approach to this particular subset of the computations.

## Standards

There are many Standards applicable to robotic systems. These range from general computer standards to embedded-specific embedded standards. Why should we be interested in standards? Standards represent an established approach, based on best practices. Standards are not created to stifle creativity or direct an implementation approach, but rather to give the benefit of previous experience. Adherence to standards implies that different parts will work together. Standards are often developed by a single company, and then adopted by the relevant industry. Other Standards are imposed by large customer organizations such as the Department of Defense, or the automobile industry. Many standards organizations exist to develop, review, and maintain standards.

Standards exist in many areas, including hardware, software, interfaces, protocols, testing, system safety, security, and certification. Standards can be open or closed (proprietary). Sometimes, the embedded systems customer will insist on adherence to specified standards.

Hardware standards include the form factor and packaging of chips, the electrical interface, the bus interface, the power interface, and others. The JTAG standard specifies an interface for debugging.

In computer architecture, the ISA specifies the instruction set and the operations. It does not specify the implementation. Popular ISA's are x86 (Intel) and ARM (ARM Holdings, LTD). These are proprietary, and licensed by the Intellectual Property holder.

In software, an API (applications program interface) specifies the interface between a user program, and the operating system. To run properly, the program must adhere to the API. POSIX is an IEEE standard for portable operating systems.

There are numerous Quality standards, such as those from ISO, and Carnegie-Mellon's CMM (Capability Maturity Model). CMM defines five levels of organizational maturity in a company or institution, and is independently audited.

Language standards also exist, such as those for the ANSI c and Java languages.

Networking standards include TCP/IP for Ethernet, the CAN bus from Bosch, and IEEE-1553 for avionics.

The ISO-9000 standard was developed by the International Standards Organization, and applies to a broad range industries. It concentrates on process. It's validation is based on extensive documentation of organization's process in a particular area, such as software development, system build, system integration, and test and certification.

It is always good to review what standards are and could be applied to an embedded system, as it ensures the application of best practices from experience, and interoperability with other systems.

The Portable Operating System Interface for Unix (POSIX) is an IEEE standard, IEEE 1003.1-1988. The standard spans some 17 documents. POSIX provides a Unix-like environment and API. Various operating systems are certified to POSIX compliance, including BSD, LynxOS, QNX, VxWorks, and others.

*ARINC 653* is a software specification (API) for space and time partitioning in safety critical real-time operating systems. Each piece of application software has its own memory and dedicated time slot. The specification dates from 1996.

# Systems Engineering Design Process

The embedded system engineering process starts with the enumeration of requirements – what is it that the system has to do? This applies at the systems level, and will be allocated to parallel paths in the software and hardware. Design methodologies involve a plan and a process for creating a system to meet requirements. The more complex the system, the more complex are the plans. The process is the key.

The design flow describes the sequence of steps in implementing the design methodology. There are various industry standard methodologies and models that have proved useful in large and complex projects.

We should apply a sound system engineering process to your project. We start by defining our requirements, which will mostly derive from the environment that we want the rover to operate in. We might be cost or time constrained, which will drive the choices for build or buy. We can keep the cost low and the schedule intact if we integrate a lot of existing parts, such as mobility platforms and small computer boards.

We can develop a set of "strawman" requirements for our Earth explorers. To these would be added the requirements from the specific environment domain the explorer has to operate in.

# Requirements

Collecting the requirements, we are defining, "What is the question?, What are we trying to do here?" We are the customer. But, before we go very far down the path of implementation, we should have a good idea what exactly it is we are trying to accomplish.

The, the requirements flow-down into more detailed levels, such as the functional requirements, the Safety requirements, the interfacing requirements, the security requirements, and such things as size, weight, power, waterproof-ness, radiation tolerance....etc. And some of these requirements may be at odds with others. Delivery time is a requirement. So is maintainability. Sometimes, even color.

The more time we spend on the requirements, and their ramifications, the easier the rest of the task will be.

What is/are out Domain(s) of operation. Are humans in the vicinity? This will drive the safety requirements.

What is the Operation duration? Do we expect the robot to be on its own for hours, days, weeks, or months? Do we anticipate unattended operation?

Will our communications be upon opportunity or continuous? What is feasible for the link? Will wifi be available, or do we need an expensive satellite modem?.

For certain underwater applications, we might use a tether.

What about ease of deployment? Do we need a truck, boat, or a helicopter to get the rover where it's going to operate?

Consider Testability - Limit complexity by design and you enhance testability. Define the testing strategy. Consider Built-In Self Test (BIST).

How much redundancy do we need and can we support? Redundancy involves extra size, weight, and power.

One we get our essential requirements together, we can define certain derived requirements from these. This would include factors such as:

Size, weight, power source, platform, mobility options, computing resources, communication resources, environmental: thermal range, moisture, radiation exposure, etc.

What support infrastructure will the rover need? How do we get it back if it get stuck, or fails in the field. For the Greenland Rover, which goes off on its own for weeks at a time, I said that if it failed, the least senior person would don a parka, grab a voltmeter, and get helicoptered to the last known location, and dropped off. We actually developed a good set of remote diagnostics and self-test routines.

# Specifications

The specifications are derived from the requirements. We need to address every requirement, but not include anything else. Every specification needs to be traceable back to a requirement, and result in a definition of what we're implementing, and a way to test that we have met the requirement.

When we do the specifications, we are answering the question,. "How will the requirements be met?" At this point we should have a complete and correct set of requirements, or there will be a lot of fixing up to do later.

We need to determine if some of the requirements are mutually exclusive? If so, we have to go back and revisit the requirements. You can't have it both ways. We need to cross-reference specifications to requirements. There are automated tools to do this for larger projects. A piece of paper works well for smaller projects.

Usage scenarios help with defining the requirements. Think about how you will use the robot platform.

Design Reviews are another tool to help us get the specifications correct, complete, and in sync with the requirements. Maybe you can get a friend or mentor to look over your project, and make suggestions. Another set of eyes is useful.

Our goals in implementation are:

- Functionality.
- Schedule: Time to deliver/time to market.
- Implement the Interface.
- Minimize Non-recurring .costs
- Minimize manufacturing cost.
- Correct Size, weight, power, energy, color, radiation tolerance.

## Design Decomposition

There may be multiple ways to implement what we want to do, so there are design trade-offs to be made. What functions are done in hardware, and which are done in software? Will this be a cpu-based design, or a custom architecture in fpga or asic. Can we use a System-on-a-chip approach? What is the interface specification – what are we talking to, and how are we talking? What are our data sources and users? How do we test and verify that the implemented system meets the requirements.

Will the system need to be changed or improved after completion? (duh!) Yes, Yes it will.

Experience counts: To the person with a hammer, all problems resemble nails. Are you more software oriented, or more hardware oriented? Mechanical?

Experience is often a major factor in implementation choices.

Our obvious design goal is to construct an implementation with the desired functionality.

A major design challenge is optimizing multiple design metrics simultaneously. Design metrics are a measurable feature of a system's implementation. Common metrics include:

- NRE cost (Non-Recurring Engineering cost): The one-time monetary cost of designing the system, including tools, both hardware and software.

- Size: the physical space required by the system

- Performance: the execution time or throughput of the system

- Cycle time – fixed or variable?

- Power: the amount of power consumed by the system (and the heat produced)

- Flexibility: the ability to change the functionality of the system easily.

- Time-to-prototype: the time needed to build a working version of the system

- Maintainability: the ability to modify the system after its initial release.

- Correctness.

- How safe is the system?

- How secure is the system?

- Ease of learning and use.

- Testability – can you get enough data to figure out what happened when things got bad?

Design metric interactions are common - improving one may worsen others. Are these unintended consequences or blessings in disguise? There are trade-offs to be made. This is, after all, an engineering project. Expertise with both software and hardware is needed to optimize design metrics.

## System Architecture

The system architecture has two major components: the software and the hardware architecture. You are free to choose either the hardware base, the software environment, either, or neither. Keep in mind, software is usually hardware dependent.

The hardware platform architecture, we'll assume for the moment, contains a cpu, memory, I/O devices, perhaps a bus to tie the I/O together. A single chip solution will give you cpu, memory, and I/O in one package, and the bus will be internal to the chip. Multicore will give you multiple cpu's to keep busy. The software architecture will be constrained by the hardware, and will be limited by considerations of performance, testability, maintenance, and experience base.

Hardware and software architectures are interlinked by considerations based on the requirements; chiefly, speed, throughput, memory size, and I/O capacity.

The hardware platform can be prototyped on an evaluation board supplied by the chip manufacturer, or third parties. It will include the classical elements of cpu, memory, and I/O, plus clock circuits, a power supply, and usually a prototype area. Sometimes, the manufacturer will make the net list of the board available, which provides a starting point for the final board design. In some cases, the prototype board can be used in the final design, if cost and schedule is an issue.

The Processor choice doesn't revolve around operations per second, or word width so much anymore. We can get a 32- r 64-bit processor with a clock in the multi-gigahertz range cheap. As we tackle more elaborate problems, even these will seem inadequate, though. We are more focused on integrated functionality, such as the number of I/O's, analog, and interrupts. We need to consider power consumption and heat generation as well.

The implementation software language is a factor, but is overshadowed by the your experience. Choose a language you are good at, and they hit the ground running.

The development platform and its requirements is a minor issue, as we can assume it runs on a pc. It may be open source, or proprietary, and there may be multiple options available.

The Real-time requirements – this is the hard part, and what differentiates embedded from desktop/server. There are two types:

Remember, with Hard Real Time requirements, things will break, and people may die if we miss the deadline. With soft real time requirements, we will make a mess, but we can recover. Most robotic systems have aspects of both hard and soft real time.

The operating system is key, and its scheduling policy is critical. Desktop operating systems, Windows, Linux, schedule fairly. Programs get equal access to the resources, and every one plays well together. Real-time operating systems are by no means fair. They have a priority scheduler, and they make sure the important tasks run, even if some lesser tasks never do. .

Power for the embedded computer can be an issue. The systems will not be plugged into the wall. It can be solar powered, or wind powered, or a combination (hybrid) system. It might have a radioisotope thermal generator, if you have really good connections. Investigate the quirks of the power source, and see how these can be addressed.

Keep in mind generated heat depends on power consumption, but battery life depends on energy consumption. Energy use is the power use, integrated in time. If we generate too much heat, we have to dump it somewhere. This might be a good thing. For a robotics project going to the Greenland Ice Shelf, we used the waste heat to keep the batteries warm. Power consumption is proportional to the voltage, squared. We might implement toggling; cycling the system on and off: more activity means more power.

We can reduce power usage by:

- Reduce power supply voltage.
- Run at lower clock frequency.
- Run at lower duty cycle
- Sleep/suspend
- Disable function units with control signals when not in use. (sleep/suspend)
- Disconnect parts from power supply when not in use.

We can employ static power management which does not depend on CPU activity, or we can let the cpu regulate itself. Dynamic power management, based on CPU activity, may be built into cpu. Keep in mind, though, that dynamic power management takes power.

So, we must determine if going into a lower power mode is worthwhile. We can model CPU power states with a power state machine model. Modern CPU's implement multiple low power modes such as sleep, doze, etc.

The system on a chip approach is particularly applicable to embedded systems. This approach involves a single chip containing a cpu, memory, and I/O, These can be purpose built for a specific application, and implemented in a ASIC or FPGA, or can be a "general purpose" or generic embedded chip. The chip can contain both digital and analog functions. The complexity of the single chip solution reflects in the lessened complexity at the board level. Fewer external interconnects leads to enhanced reliability. But, it is important to realize that the system complexity does not disappear. It is at the box, board or chip level.

Fault Tolerant Design

In this design approach, a system is designed to continue to operate properly in the event of one or more failures. It is sometimes referred to as graceful degradation. There is, of course, a limit to the number of faults or failures than can be handled, and the faults or failures may not be independent. Sometimes, the system will be designed to degrade, but not fail, as a result of the fault. Fault recovery in a fault-tolerate design is either roll forward, or roll back. Roll back refers to returning the system state to a previous check-pointed state. Roll forward corrects the current system state to allow continuation. Can you recover your platform if it fails, or will it be at the bottom of a lake, or achieving terminal velocity as it falls through the atmosphere (toward your neighbor's dog house)?

Redundancy

Redundancy refers to the technique of having multiple copies of critical components. This can refer to hardware or software. This increases the reliability of the system. Redundant units can be deployed in parallel, such as extra structural members, where each single unit can handle the load. This provides what is referred to as a margin of safety. An improvement in reliability can be achieved by simply adding a second unit in many cases.

In certain systems that are responsible for safety-critical tasks, we might triplicate the critical portion, which, reduces the probability of system failure to small, acceptable, levels.

Of course, if there is a common error in the three units, we have not increased our reliability. This situation is referred to as a common mode or single point error. Another problem is in the voting logic, that makes the decision that an error has been made, and switches controllers. At least one satellite launch failed because the voting logic made the wrong choice. Redundancy carries penalties in size, weight, power, cost, and testing complexity.

Fault isolation allows the system to operate around the failed component, using backup or alternative modules. Fault containment strives to isolate the fault, and prevent propagation of the failure.

Systems can be designed to be fail-safe, fail-soft, or can be "melt-before-fail." the more fault tolerant that is built into a system, the more it will cost, and the more difficult it will be to test. It is important not to increase the complexity to the point where the system is not testable, and is "designed to fail."

## Safety

Most of what we need to know about Robot Safety can be found in Asimov's 3 laws.

Mobile Robotic systems operate in the real world, and the real world can be scary. We need to be aware of the hazards that a mobile robot systems can present to others, and the hazards it itself can be subject to. We will cover some of those in the section on security. A good starting point for robotic safety comes from a science fiction book published in 1942 by Isaac Asimov. In his short story, *Runaround*, he introduced his *Three Laws of Robotics,* which have stood the test of time. From their introduction in speculative fiction to their influence on industrial systems, they are well-thought-out.

And, they are:

- A robot may not injure a human being or, through inaction, allow a human being to come to harm.

- A robot must obey the orders given to it by human beings, except where such orders would conflict with the First Law.

- A robot must protect its own existence as long as such protection does not conflict with the First or Second Law.

Asimov went on to write many robotics stories, where the effect of the three laws were seen in some unusual situations. He actually attributes the formulation of his laws as a conversation with John Campbell in 1940. Asimov always assumed the robots he wrote about had inherent safeguards.

So, based on Asimov's laws as a starting point, we can derive some requirements for our robotic systems. First, to not harm a human, the robot must have passive and active safety systems. It must be aware of humans within its reach or task space. Speaking as one who was pinned to a wall by a 350 pound robot cart, a human-sensor is a good idea. If you are operating your quadcopter, it is not a good idea to fly it into another person (dog, car...). The flow-down safety from the 3-laws continue. Consider safe design, and safe operation at the beginning.

One approach for robot safety is to use a separate safety watchdog computer system, with a separate low level hardwire control. The separate safety watchdog computer usage did not imply that it is the sole repository of safety responsibility. The implementation of safety was distributed in the system, from the workstation to the robot control computers. At all levels, the system checked the "reasonableness" of actions before they are carried out. Each safety computer monitored both control computers, both joint controllers, and both manipulators. The safety computers were implemented as a redundant pair, and either could safe the system in a problem situation.

The safety system can be a fail-safe system, and the main controller can be less than bulletproof. The implementation of the true separation of inhibits and monitoring of inhibits was carefully examined, to ensure that "sneak paths" did not exist for unintentionally removing functional inhibits.

## Security

Are you familiar with the term, robo-jacked? Well, I just made that up. But, it refers to a situation in which some one else takes over your remote robot.

All embedded systems have aspects of security. Embedded systems on robots operate in an unfriendly world. They are vulnerable to variations of hacking, viruses and malware, theft, damage, spoofing, and other nasty techniques from the desktop/server world. GPS systems can be hacked to provide incorrect location or critical time information. Cell phones and tablets are connected wirelessly to large networks. A bored teenage hacker in Europe took over the city Tram system as his private full-scale railroad, using a TV remote. What about the teenager in an internet café is a third-world country. Can he take over and play with your robot?

Some of these issues are addressed by existing protocols and standards for access and communications security. Security may also imply system stability and availability. Standard security measures such as security reviews and audits, threat analyses, target and threat assessments, countermeasures deployment, and extensive testing apply to the embedded domain.

A security assessment of a system involves threat analysis, target assessment, risk assessment, countermeasures assessment, and testing. This is above and beyond basic system functionality.

The completed functional system may need additional security features, such as intrusion detection, data encryption, and perhaps a self-destruct capability. Is that self-destruct capability secure, so not just anyone can activate it? All of these additional features use time, space, and other resources that are usually scarce in embedded systems.

Virus and malware attacks on desktops and servers are common, and an entire industry related to detection, prevention, and correction has been spawned. These issues are not as well addressed in the embedded world. Attacks on new technology such as cell phones, pda's, tablets, and GPS systems are emerging. Not all of the threats come from individuals. Some are large government-funded efforts or commercial entities seeking proprietary information or market position. Security breaches can be inspired by ideology, money, or fame considerations. The *CERT* (Computer Emergency Response Team) organization at Carnegie Mellon University, and the *SANS Institute* (SysAdmin, Audit, Networking, and Security) track security incidents.

Techniques such as hard checksums and serial numbers are one approach to device protection. Access to the system needs to be controlled. If unused ports exist, the corresponding device drivers should be disabled, or not included. Mechanisms built into the cpu hardware can provide protection of system resources such as memory.

Security has to be designed in from the very beginning; it can't just be added on. Memorize this. There will be a quiz.

Even the most innocuous embedded platform can be used as a springboard to penetrate other systems. It is essential to consider security of all embedded systems, be aware of industry best practices and lessons learned, and use professional help in this specialized area.

The first detection of *backdoor code* in a military grade FPGA chip came in May of 2012. This was detected in an Actel ProASIC3 chip. It was built into the silicon and was activated by a secret key code. This caused much distress worldwide in the FPGA/ASIC world, and for their military customers. Although this was the first detected instance of this security breach, it was probably not the first instance. The story of how it was discovered is of interest. We can expect more of this type of behavior in the future of embedded systems.

Ref:
http://www.cl.cam.ac.uk/~sps32/Silicon_scan_draft.pdf

# Wrap-up

We are on the cusp of a new great era of exploration, of the most important planet we know. You can participate in this through your surrogate robot. It is becoming easier and cheaper all the time.

Keep in mind you are working in a cooperative ecosystems of developers and implementers world-wide. You can use the work of others to build upon, and you should pay back to the community.

Think of a model where you would like to explore the Great Barrier Reef off Australian, but it's too long a plane ride, you never learned to scuba, and actually, you can't swim. Send your underwater robot, with the hi-res video. Telecommand it via an Ap on your smartphone.

Join an expedition in Belize to uncover previously unknown MesoAmerican ruins. From your armchair, pilot the mapping drone, looking for areas of interest. Mark these and send the GPS co-ordinates to the field team. No pesky mosquitoes to deal with.

Could you be the one to find Genghis Khan's tomb, from your hot tub? Do you want to explore Antarctica, but are confined to a wheelchair? No problem. The technology to do all these things exists right now, and is getting cheaper every day. At some point, you will be able to command a rover on the Moon, or Mars. Go for it!

# Getting involved; getting students involved

There are many programs to get students involved in robots projects such as rovers. If you're up to speed in this field, you might want to consider volunteering at a local high school or college. If you're in it for the fun, go for it. If you're just starting out, there are some choices to work with and learn from others.

STEM

STEM stands for science, technology, engineering, and mathematics. The STEM fields are those academic and professional disciplines that fall under the umbrella areas represented by the acronym. According to both the United States National Research Council (NRC) and the National Science Foundation (NSF), the fields are collectively considered core technological underpinnings of an advanced society. In many forums (including political/governmental and academic) the strength of the STEM workforce is viewed as an indicator of a nation's ability to sustain itself.

The Science, Technology, Engineering, and Mathematics Education Coalition works to support STEM programs for teachers and students at the U. S. Department of Education, the National Science Foundation, and other agencies that offer STEM related programs.

# FIRST

FIRST (For Inspiration and Recognition of Science and Technology) is an organization founded by inventor Dean Kamen in 1989 to develop ways to inspire students in engineering and technology fields. The organization is the foundation for the FIRST Robotics Competition, FIRST LEGO League, Junior FIRST LEGO League, and FIRST Tech Challenge competitions.

The FIRST® LEGO® League is an international competition organized by FIRST for elementary and middle school students. Each year, a new challenge is announced that focuses on a different real-world topic related to the sciences. The robotics part of the competition revolves around designing and programming LEGO Robots to complete tasks. The students work out solutions to the various problems they are given and then meet for regional tournaments to share their knowledge, compare ideas, and display their robots. FIRST LEGO League is a partnership between FIRST and the LEGO Group. It also has a scaled-down robotics program for children ages 6–9 called Junior FIRST LEGO League.

NASA/GSFC Summer Engineering Robotics Boot Camp

This program ran for eight years under the guidance of Mike Comberiate (*NASA-Mike*). It brought undergraduates, graduate students, even some high-schoolers together for an intense 8-week session at

Goddard. It was an expansion of the 25-year program COOL-SPACE, Communications Over Obscure Locations, Special Purpose Advanced Communications Equipment, which provided INTERNET access to Antarctica via satellite. In the 2011 session, there were over 800 applicants, down-selected to about 40. International students were accepted. The level of energy and achievement is high. Recent projects have involved tracked vehicles in Antarctica in 2008, an autonomous vehicle for Greenland ice sheet exploration, and an autonomous meteorite finder for Antarctica. At the conclusion of the summer program, it is not unusual for students to be hired by NASA or a Contractor. The Program was targeted at STEM careers, with multi-discliplinary engineering.

Zero Robotics Competition

This program involves a series of robots already on the International Space Station called SPHERES (Synchronized Position Hold, Engage, Reorient Experimental Satellites). These have a mass of around ten pounds, and a diameter of 8 inches. They use twelve $CO_2$ thrusters for movement, and are battery powered. They were developed at the MIT Space systems Laboratory as a testbed for control, autonomy, and metrology for distributed spacecraft and docking missions. The SPHERES were inspired by the Training Remotes from the Star Wars films. There are three SPHERES, in different colors.

As a team, they can control their relative their relative position and orientation. They have been tested aboard KC-135 aircraft flying zero-gravity flight paths (the Vomit Comet), and were delivered to the International Space Station (ISS) in 2006.

The NASA/MIT Competition allows teams to develop software for the SPHERES, and test it in a simulation environment. Selected teams test their software on SPHERES in an air-bearing floor facility. In December 2011, a few teams tested their code and algorithms in the SPHERES onboard the ISS.

On your own

Here are some suggested approaches to inexpensive personal robot projects you can do on your own. Also, check local high schools and colleges for robotics clubs and programs. If you are experienced, volunteer as a mentor. If you are starting out new, it is good to work with a group of like-minded individuals.

Let me explain some approaches, that may make it easier for you. First, consider whether you are a scientist, or an engineer. A scientist seeks out new knowledge, and an engineer likes to build things. The scientist uses the Scientific Method, and the engineer relies on good System Engineering practice. Both of these approaches work well in their own domains.

Ok, do you want to build a robot so you can use it to explore some domain of interest, or do you have a domain of interest in mind you want to explore, and a robot platform is the best way to do it? Are you an engineer or a scientist? I are an engineer.

There is a difference of approach in the two domains, but neither is right or wrong – they are synergistic. The scientist often asks the engineer to build the hardware and software, so he/she can focus on the problem. The engineer often needs the scientist to pose the problem for which the hardware/software needs to be built.

Can you be both? Certainly. You won't be as good at either, but it works. Most scientists are hands-on enough to put together hardware and software, but most also prefer to focus on the problem. Engineers are overjoyed to be presented with a project to build something that hasn't been done yet.

For our Greenland Rover Project, the payload was a unique ice-penetrating radar, developed by a scientist expert in such things. We integrated it on the Rover system. It was.....a radar unit built by a scientist (apologies, Hans-Peter). It worked. An engineer wouldn't have built it that way, but we were smart enough not to touch it.

Let me explain the distinction from a NASA perspective. In the early days of satellites, each was built from scratch, and each was different. It was quickly realized, however, that the same platform could support a variety of instruments. The commonality was the power system, the structure, the command and telemetry, and the attitude control. This lead to a common design, where the engineering part ("the bus") was the same, and the instruments differed. There was a defined interface, and defined services that the bus provided to the instruments. The bus didn't care about the instruments. They just provided data that was a pass-through. The bus could support missions both by the Earth scientists (they look down) and the astrophysicists (they look up). There was an economy of scale in reusing the design, and in support of missions.

Let me mention is passing the importance of mathematics. Both groups, engineers and scientists, use a lot of math. Scientists like it; actually, mathematicians really like it, and engineers use it as a tool Take all the math you can; do all the hard problems at the end of the chapters.

Where do software developers fit it? Both scientists and engineers can develop software, and mathematicians are good at it as well. Software can be an art form, and a tool. Develop your skill with tools of all kinds.

So, essentially, are you interested in the robot for its own sake (the bus), or in what it can carry and do? Both approaches are valid.

Choose your domain:

Land, water, air, underwater. If you want high altitude, look into balloons and blimps. If you want a more expensive option, take a look at the cube sat program.

Radio controlled "toys" in all of these areas provide a ready-made platform, allowing you to focus on the smarts of the system.

Develop your expertise. Get a small embedded board, and learn to develop code for it, and interface sensors and motors. Get it working on the workbench.

Integrate your electronics with your platform, and go explore your domain of interest. Write up what you do, and share back with the community. Help those staring out. Enjoy the project. Go for it!

Some Suggested Projects

Engineering

Multiple, cooperating systems. Wheeled or tracked platform that can deploy a smaller wheeled/tracked platform, and can retrieve it later. Ground platform that can deploy and retrieve an aerial platform. Water-borne platform that can deploy and retrieve a submersible. Amphibious platform that can operate well on land and in the water.

Science

Autonomous water sampling from large bodies of water such as lakes and bays. The system should be able to take samples from various depths, and return them for analysis.

Remotely deployed weather station. It can be airdropped and mobile. Returns temperature, humidity, and wind data on the periphery of a forest fire area.

Flying platform, mid-altitudes, to collect and return air samples for lab analysis. Can be launched downwind of suspected polluters. Can also collect wind speed and direction information for potential wind farm sites.

# Glossary

Aerostat – flight system deriving lift from buoyancy.
Aerobot – aerial robot
Agent – an autonomous entity (hardware/software) which is goal-seeking.
AHS – autonomous haulage system.
AI – artificial intelligence.
ALS – autonomous loading system.
ANTS – Autonomous Nano Technology Swarm
Api – applications program interface
Arduino – small embedded systems architecture.
ASIC – application specific integrated circuit. Specific hardware.
Atmosat – platform operating at high altitude in the atmosphere, for extended periods.
AUV - autonomous underwater vehicle
BIOS – Basic I/O system – initialization.
Bit – smallest unit of binary information. Two states.
Bluetooth – a short range radio standard for data.
BSD – Berkeley Systems Distribution (of Unix)
BSP – board support package – customization code for specific hardware.
Byte – collection of 8 bits.
C – a programming language.
CARACaS (U. S. Navy) - Control Architecture for Robotic Agent Command and Sensing
CAST – (CalTech) Center for Autonomous Systems & Technologies.

Cubesat – small satellite that can be developed by schools or individuals (standard)
CMU – Carnegie Mellon University.
Codec – coder/decoder
copyleft – license for open source software
CoSTAR - Collaborative SubTerranean Autonomous Robots
Cpu – central processing unit.
Cubesat – small satellite that can be developed by schools or individuals (standard)
Dalvik – the virtual machine in Android.
DARPA – Defense Advanced Research Projects Agency.
Drone – unmanned aerial vehicle.
DSP – digital signal processor
DSN – NASA's Deep Space Network
duty cycle – percentage of "ON" in an on-off cycle.
eeprom – electrically erasable programmable read-only memory. Mostly superseded by flash.
Eloran – enhanced LORAN navigation aid.
Ethernet – a networking protocol; wired or wireless.
FPGA – field programmable gate array.
Gbytes - $10^9$ bytes.
GHz - $10^9$ hertz
Gpio – general purpose input output
GPS – global positioning system, series of navigation satellites.
Grover – Greenland Rover.
GSGC – Goddard Space Flight Center, lead NASA Center for Earth observation.
HDMI – High Definition Multimedia Interface

Hexcopter - a small aircraft with six small horizontal rotors, like a helicopter.
Homeostasis – a self-monitoring, self-regulating system. System.
IDE – Integrated Development Environment (toolset).
IMU – inertial measurement unit.
IoT – Internet of Things.
Iridium – a satellite system for global communications.
Isa – instructure set architecture.
Java – a programming language.
Kaist - Korea Advanced Institute of Science and Technology.
Kbytes – $10^3$ bytes.
Lidar – radar, using light. Same thing, different frequency.
LDC – LORAN data channel.
Linux – open source operating system; unix-like
LioP – lithium polymer battery technology.
LORAN – hyperbolic radio navigation system, U. S. Coast Guard.
Lorax – Life on Ice Robotic Antarctic Explorer – CMU project.
Lunokhod – Russian lunar probe; translates as "moonwalker"
LUT – (Sweden) Lulea University of Technology.
Malware – malicious software.
Mbytes – $10^6$ bytes.
MEMS - microelectronic mechanical systems – producing mechanical systems such as gyros using microelectronics, fabrication technology.
MER – Mars Exploration Rover.

Metadata – data about the data. Date, time, modified? Etc.
microcontroller – microprocessor plus memory and I/O.
mips – millions of instructions per second.
MIT – Massachusetts Institute of Technology
MSL – Mars Science Lab.
Multicore – computer architecture with multiple processors on one chip.
Mutex – a mutual exclusion mechanism in hardware (traffic light) or software.
NASA – National Aeronautics and Space Administration.
NOAA – National Oceanographic Atmospheric Administration.
NREC – National Robotics Engineering Center, CMU.
NSF – National Science Foundation.
ONR – (U.S.) Office of Naval Research.
Pc – personal computer
pda – personal digital assistant
Perigee – point of closest approach to Earth in a satellite orbit.
PWM – pulse width modulation; used for dc motor speed control.
Python – programming language; large man-eating snake.
Quadcopter – a small aircraft with four small horizontal rotors, like a helicopter.
Regolith - layer of unconsolidated rocky material covering bedrock.
Rov – remotely operated vehicle.
Sata – a serial disk interface standard.
SatCom – satellite communications.

Seu – single event upset in electronics, due to radiation.
SMART – Super Miniaturized Addressable Reconfigurable Technology.
Sol – local day.
Stiction – static friction.
Uart – universal asynchronous receiver transmitter.
UAS – unmanned aerial system.
UaV – unmanned aerial vehicle, or remotely piloted aircraft, or drone.
UGV – unmanned ground vehicle.
Unix – operating system from Bel Labs, written in the c language.
URL – uniform resource locater. Used as a reference to a resource on the Internet.
Usb – universal serial bus, a communications standard.
UUV – unmanned underwater vehicle.
Vga – a video display standard.
VxWorks – a real time operating system from Wind River.
WPS – wifi positioning system.147 pages
WiFi – short range radio-based networking.

# Bibliography

Abut, Huseyin (ed.) et all *Advances for In-Vehicle and Mobile Systems: Challenges for International Standards* Springer; 1 edition, 2007, ISBN-1 038733503X.

Afanassieff, Jean *Tank on the Moon, DVD, 2007.*

*Antonelli, Gianluca Underwater Robots, Springer; 3rd ed. 2014, ISBN- 3319028766.*

Albus, James S. *System Description and Design Architecture for Multiple Autonomous Undersea Vehicles,* University of California Libraries, January 1, 1988, ISBN-10: 1125517441.

Albus, James A. *Brains, Behaviour and Robotics*, McGraw-Hill Inc., 1st Edition, December 1, 1981, ISBN-10: 0070009759.

Anas, Brittany, "Researchers to Fly Unmanned Planes over Greenland," July 16, 2008, Scripps News.

Armandt, David *Controlling robotic swarm behavior utilizing real-time kinematics and artificial physics,* 2016, ASIN-B07Q24X6K2.

Baker, et al "A Campaign in Autonomous Mine Mapping," Robotics Institute, Carnegie Mellon University, 2004.

https://www.ri.cmu.edu/pub_files/pub4/baker_christopher_2004_1/baker_christopher_2004_1.pdf

Berns, Karsten; Puttkamer, Ewald; Vieweg, M. L.; *Autonomous Land Vehicles*, Teubner; 2009 edition, ISBN-10: 3834804215.

Bloydyk, Gerardus *Swarm Robotics Platforms A Complete Guide - 2020 Edition,* ASIN - B084V76PRS.

Campbell, Brenton *Human robotic swarm interaction using an artificial physics approach,* 2014, ASIN - B07PPXQSZZ.

Bräunl, Thomas *Embedded Robotics: Mobile Robot Design and Applications with Embedded Systems* Springer; 2nd ed., 2006, ISBN- 3540343180.

Castellanos ,Jose A.; Tardós, Juan D. *Mobile Robot Localization and Map Building: A Multisensor Fusion Approach,* Springer; 2000 ed , 2000, ISBN- 0792377893.

Castelvecchi, Davide "Invasion of the Drones: Unmanned Aircraft Take Off in Polar Exploration, March 2010, Scientific American.

Christ, Robert D.; Wernli Sr. Robert L. *The ROV Manual: A User Guide for Observation Class Remotely Operated Vehicles,* Butterworth-Heinemann; 1st ed, 2007, ISBN-0750681489 .

CMU "A Robot for Volcano Exploration," www.robovolc.dees.unict.it Carnegie Mellon University, Field Robotics Center, www.frc.ri.cmu.edu

Cook, Gerald *Mobile Robots: Navigation, Control and Remote Sensing,* Wiley-IEEE Press; 1st ed, 2011, ISBN-0470630213.

de Croon, G.C.H.E;. Perçin M; Remes, B.D.W.; Ruijsink, R.;de Wagter, C. *The DelFly: Design, Aerodynamics, and Artificial Intelligence of a Flapping Wing MAV* Springer; 2014, ISBN- 9401792070.

DeMuth, Brian and Eisenreich, Dan *Designing Embedded Internet Devices*, Newnes, 2002, ISBN 1878707981.

Dudek, Gregory; Jenkin, Michael *Computational Principles of Mobile Robotics* Cambridge University Press; 2nd ed, 2010, ISBN-0521692121.

Alex Ellery, "Planetary Rovers Robotic Exploration of the Solar System," 2016, Springer Link, ISBN: 978-3-642-03259-2.

Enzmann, Robert Duncan " Unmanned Exploration of Planetary Surfaces," Annals of the New York Academy of Sciences, Volume 163, Second Conference on Planetology and Space Mission Planning, pp 387–395, September 1969.

Everett, H. R. *Sensors for Mobile Robots*, 1995, CRC Press, ISBN 1568810482.

Everett, Hobart R. "A Microprocessor Controlled Autonomous Sentry Robot," October 1982, Thesis, Navel Postgraduate School, Monterey, CA, A125239.

Fischer-Cripps, Tony *Newnes Interfacing Companion: Computers, Transducers, Instrumentation and Signal Processing* Newnes (December 20, 2002), ISBN 0750657200.

Floreano, Dario (Ed); Zufferey, Jean-Christophe (Ed); Srinivasan, Mandyam V.(Ed); Ellington, Charlie (Ed) *Flying Insects and Robots,* Springer, 2009, ISBN-354089392X.

Ge, Shuzhi Sam *Autonomous Mobile Robots: Sensing, Control, Decision Making and Applications,* CRC Press, 2006, ISBN-0849337488.

General Electric, *Embedded Computing Technologies for Unmanned Vehicles*, 2010, GE Intelligent Platforms, www.ge-ip.com

Geologic Surveys: Mineral Exploration at the Ends of the Earth, Mining-Technology.com, Nov. 2010.

Graham, Brad; McGowan, Kathy *Build Your Own All-Terrain Robot* McGraw-Hill/TAB Electronics; 1st ed, 2004, ISBN-007143741X.

Hebert, Martial H. (Ed); Thorpe, Charles E. (Ed); Stentz, Anthony (Ed); *Intelligent Unmanned Ground Vehicles: Autonomous Navigation Research at Carnegie Mellon Springer; 1997 ed,, 1996, ISBN-0792398335.*

Iagnemma, Karl; Dubowsky, Steven *Mobile Robots in Rough Terrain: Estimation, Motion Planning, and Control with Application to Planetary Rovers, Springer, ISBN-10: 3642060269.*

Jaulin, Luc *Mobile Robotics,* 1st Edition, ISTE Press - Elsevier; 2015, ISBN-1785480480.

Jaulin, Luc (ed); Le Bars, Fabrice (ed), *Robotic Sailing 2013: Proceedings of the 6th International Robotic Sailing Conference,* Springer; Softcover reprint of the original 1st ed. 2014 edition ISBN-10: 3319033794.

Kelly, Alonzo *Mobile Robotics, Cambridge University Press; 1st edition, 2014, ASIN: B00E99YN9C.*

Jones, Joseph L; Seiger, Bruce A.; Flynn, Anita M. *Mobile Robots: Inspiration to Implementation, Second Edition, A K Peters/CRC Press; 2nd ed, 1998, ISBN-1568810970.*

Kirianaki, Nikolay V. et al *Data Acquisition and Signal Processing for Smart Sensors,* 2002, Wiley, ISBN 0470843179.

Kortenkamp, David (Ed); Bonasso, R Peter (Ed), Murphy, Robin R. (Ed) *Artificial Intelligence and Mobile Robots: Case Studies of Successful Robot System* AAAI Press; 1st edition, 1998, ISBN-0262611376.

Lakdawalla, Emily "The Design and Engineering of Curiosity: How the Mars Rover Performs its Job", 2018, Springer Praxis, ISBN-978-3319681443.

Leary, Warren E. "Robot Named Dante To Explore Inferno Of Antarctic Volcano." New York Times. December 8. 1992.

Leary, Warren E. "Robot Completes Volcano Exploration," New York Times, August 3, 1994.

Leary, Warren E., "Robot Is Nearing Goal Inside Active Volcano," New York Times, August 2, 1994.

Leary, Warren E., "Hunt for Meteorites In Antarctica Enlists a Novel Recruit," New York Times, January 18, 2000.

Leary, Warren E. "Hardier Breed of Antarctic and Lunar Explorers: Robots," New York Times, May 13, 1997.

Leveson, Nancy G. *System Safety and Computers*, Addison-Wesley, 1995, ISBN: 0-201-11972-2.

Mahaney, W. C. et al, "Morphogenesis of Antarctic Paleosols: Martian Analogue," Icarus, Volume 154, Issue 1, November 2001, Pages 113-130.

Manoj Franklin, Zahran, Mohamed *Single-Chip Parallel Processing: The Era of Multicores and Manycores,* Morgan Kaufmann, 2011, ISBN-10: 0123744970.

Meystel, Alexander M.; Albus, James S.; *Intelligent Systems: Architecture, Design, and Control* (Wiley Series on Intelligent Systems), Wiley-Interscience; 1st edition, October 10, 2001, ISBN-10: 0471193747.

Moore, Steven W.; Bohm, Harry; Jensen, Vicky *Underwater Robotics: Science, Design & Fabrication,* Marine Advanced Technology Edu; 1st ed, 2010, ISBN-0984173706.

Nonami, Kenzo; Kendoul, Farid; Suzuki, Satoshi; Wang, Wei; Nakazawa; Daisuke *Autonomous Flying Robots: Unmanned Aerial Vehicles and Micro Aerial Vehicles* Springer; 2010 edition, 2010, ISBN- 4431538550.

Nourbakhsh, Illah Reza; Scaramuzza, Davide; Siegwart, Ronald *Introduction to Autonomous Mobile Robots*, 2nd edition, 2011, TBS, 2011, ISBN-8120343220.

Pfister, Cuno *Getting Started with the Internet of Things: Connecting Sensors and Microcontrollers to the Cloud,* O'Reilly Media; 1st edition, June 2, 2011, ISBN-1449393578.

Rice, Doyle, "Unmanned Research Robots Destined to Roam Ice Sheets," Nov. 24, 2008, USA Today.

Rizzi, Alfred A.; Whitcomb, Louis L.; Koditschek, Daniel E. "Distributed Real-Time Control of a Spatial Robot Juggler", Computer, May 1992 v 25 n 5 p 12.

Seto, Mae L. (Ed) *Marine Robot Autonomy,* Springer; 2013 edition, 2012, ISBN- 1461456584.

Siegwart, Roland and Nourbakhsh, Illah R. *Introduction to Autonomous Mobile Robots,* The MIT Press, 2004, ISBN- 026219502X.

Spaanenburg, Lambert and Spaanenburg, Hendrik, *Cloud Connectivity and Embedded Sensory Systems,* Springer; 1st Edition, 2010, ISBN-1441975446.

Shuzhi, Sam Ge, *Autonomous Mobile Robots: Sensing, Control, Decision Making and Applications* (Automation and Control Engineering), CRC Press; 1st edition, May 4, 2006, ISBN-10: 0849337488.

Stakem, Patrick H. *Robots Underground,* 2022, ISBN-

Stakem, Patrick H. "The Brilliant Bulldozer: Parallel Processing Techniques for Onboard Computation in Unmanned Vehicles", 15th AUVS Symposium, San Diego, Ca. June 6-8, 1988.

Truszkowski, Walt *Autonomous and Autonomic Systems: With Applications to NASA Intelligent Spacecraft Operations and Exploration Systems,* Springer; 1st Edition. edition, 2009, ISBN-1846282322.

Truszkowski, Walt; Clark, P. E.; Curtis, S.; Rilee, M. Marr, G. *ANTS: Exploring the Solar System with an Autonomous Nanotechnology Swarm.* J. Lunar and Planetary Science XXXIII (2002)

Tzafestas, Spyros G. *Introduction to Mobile Robot Control*, Elsevier; 1st ed, 2013, ASIN: B00G4N7JLA, ISBN: 0124170498.

Tzafestas, Spyros G. *Advances in Intelligent Autonomous Systems,* Kluwer Academic; 1999 ed , 1999, ISBN-0792355806.

Usher, M. J. and Keating, D. A. *Sensors & Transducers: Characteristics, Application., Instrumentation & Interfacing* 1996, Scholium International; 2nd edition, ISBN 0333604873.

Wadoo, Sabiha; Kachroo, Pushkin *Autonomous Underwater Vehicles: Modeling, Control Design and Simulation,* CRC Press, 2010, ISBN-1439818312.

Wolf, Wayne, *High-Performance Embedded Computing: Architectures, Applications, and Methodologies*, Morgan Kaufmann, 2006, ISBN- 978-0123694850.

Yamasaki, H. (ed) *Intelligent Sensors*, 1996, Elsevier Science, ISBN 0444541543.

NASREM and RCS

Albus, J.S.; Lumia, R.; McCain, H. "Hierarchical Control of Intelligent Machines Applied to Space Station Telerobots, IEEE Transactions on Aerospace and Electronic Systems, Sept 1988, V 24 n 5 pp 535-541.

Albus, James S.; McCain, Harry G.; Lumia, Ron "NASA/NBS Standard Reference Model for Telerobot Control System Architecture (NASREM)," NIST Technical Note 1235, 1989 Ed.

Albus, James, et al, 4D/RCS: A Reference Model Architecture for Unmanned Vehicle systems, Version 2, NISTIR 6910, NIST, August 2002, http://www.isd.mel.nist.gov/documents/albus/4DRCS_ver2.pdf.

"Servo Level Control for Manipulation in the NASREM Architecture," Real Time Control Group, Robot Systems Division, NBS (NIST), June 8, 1987.

Fiala, John "Manipulator Servo Level Task Decomposition," ICG-#002, NIST, Dec. 3, 1987.

Fiala, John "Interfaces to Teleoperation Devices," ICG-#004, Dec. 3, 1987.

Gazi, Veysel, Moore, Mathew L., Passino, Kevin M., Shakleford, William P., Proctor, Frederick M., Albus, James S. *The RCS handbook: Tools for Real Time Control Systems Software Development,* 2001, Wiley Interscience ISBN 0471435651.

Kagan, Eugene, Shvalb, Nir, *Autonomous Mobile Robots and Multi-Robot Systems: Motion-Planning, Communication, and Swarming,* 2019, ASIN-B07XC9CMFN.

Norton, Charles D. *Sensor Web Technology Challenges and Advancements for the Earth Science Decadal Survey Era,* 2011, ASIN-B01D54HP1Y.

Stakem, Pat, Lumia, Ron, Smith, Dave, "A Computer and Communications Architecture for the Flight Telerobotic Servicer," June 24, 1988, ICG-#20, Intelligent Controls Group, Robot Systems Division, National Bureau of Standards.

Stakem, Patrick H. *Earth Rovers: for Exploration and Environmental Monitoring,* 2014, PRRB Publishing, ISBN-152021586X.

*Stakem, Patrick H. Swarm Robotics,* ISBN-979-8534505948.

Stakem, Patrick H. *Mobile Cloud Robotics,* 2018, PRRB Publishing, ISBN- 1980488088.

Sze-Tek, Terence Ho *Investigating ground swarm robotics using agent based simulation,* 2006, ASIN - B07MCMGN3S.

Tarnoff, Nicholas Jacoff, Adam Lumia, Ronald, "World Model Registration for Effective Off-Line Programming of Robots," NIST, Jan. 1990 (3$^{rd}$. International Symposium on Robotics and Manufacturing (ISRAM) Vancouver, B.C. Canada July 18-20, 1990.

Truszkowski, Walt, et al *Autonomous and Autonomic Swarms,* https://ntrs.nasa.gov/search.jsp?R=20050210015 2020-02-09T20:09:51+00:00Z

Wallace, Luke; Lucieer, Arko; Watson, Shristopher; Turner, Carren *Development of a UAV-Lidar System with Application to Forrest Inventory*, Remote Sens. 2012, 4, 1519-1543; www.mdpi.com/journal/remotesensing

Wavering, Albert J. "Manipulator Primitive Level Task Decomposition," ICG-#003, Jan. 5, 1988.

# Resources:

http://spectrum.ieee.org/automaton/robotics/military-robots/navy-drone-boat-swarm-practices-harbor-defense

http://www.afr.com/news/world/north-america/in-the-next-step-in-warfare-the-us-develops-underwater-drones-20161125-gsxdw1

Morris, Aaron "Recent Developments in Subterranean Robotics, 2005, https://www.ri.cmu.edu/pub_files/pub4/morris_aaron_christopher_2006_1/morris_aaron_christopher_2006_1.pdf
http://www.cbsnews.com/news/underwater-drones-openrov-trident-citizen-scientists-exploration/

"Groundhog Debut, Robot Successfully Maps Portion of Abandoned Mine,"
https://www.cmu.edu/cmnews/extra/2002/021031_groundhog.html

"Groundhog Debut, Robot Successfully Maps Portion of Abandoned Mine,"
https://www.cmu.edu/cmnews/extra/2002/021031_groundhog.html

http://dronecenter.bard.edu/underwater-drones-updated/

http://electronics360.globalspec.com/article/6756/six-underwater-drones-making-a-technology-splash

http://www.providencejournal.com/news/20160916/underwater-drones-collect-key-data-on-hermine

openrov.com

https://www.kickstarter.com/projects/1359605477/fathom-one-the-affordable-modular-hd-underwater-dr/community

https://costar.jpl.nasa.gov/

https://www.subtchallenge.com/SubTv.html

https://mars.nasa.gov/mer/mission/technology/autonomous-planetary-mobility/

# If you enjoyed this book, you might also be interested in some of these.

Stakem, Patrick H. *16-bit Microprocessors, History and Architecture*, 2013 PRRB Publishing, ISBN-1520210922.

Stakem, Patrick H. *4- and 8-bit Microprocessors, Architecture and History*, 2013, PRRB Publishing, ISBN-152021572X,

Stakem, Patrick H. *Apollo's Computers,* 2014, PRRB Publishing, ISBN-1520215800.

Stakem, Patrick H. *The Architecture and Applications of the ARM Microprocessors,* 2013, PRRB Publishing, ISBN-1520215843.

Stakem, Patrick H. *Earth Rovers: for Exploration and Environmental Monitoring,* 2014, PRRB Publishing, ISBN-152021586X.

Stakem, Patrick H. *Embedded Computer Systems, Volume 1, Introduction and Architecture*, 2013, PRRB Publishing, ISBN-1520215959.

Stakem, Patrick H. *The History of Spacecraft Computers from the V-2 to the Space Station*, 2013, PRRB Publishing, ISBN-1520216181.

Stakem, Patrick H. *Floating Point Computation*, 2013, PRRB Publishing, ISBN-152021619X.

Stakem, Patrick H. *Architecture of Massively Parallel Microprocessor Systems*, 2011, PRRB Publishing, ISBN-1520250061.

Stakem, Patrick H. *Multicore Computer Architecture,* 2014, PRRB Publishing, ISBN-1520241372.

Stakem, Patrick H. *Personal Robots*, 2014, PRRB Publishing, ISBN-1520216254.

Stakem, Patrick H. *RISC Microprocessors, History and Overview,* 2013, PRRB Publishing, ISBN-1520216289.

Stakem, Patrick H. *Robots and Telerobots in Space Application*s, 2011, PRRB Publishing, ISBN-1520210361.

Stakem, Patrick H. *The Saturn Rocket and the Pegasus Missions, 1965,* 2013, PRRB Publishing, ISBN-1520209916.

Stakem, Patrick H. *Visiting the NASA Centers, and Locations of Historic Rockets & Spacecraft,* 2017, PRRB Publishing, ISBN-1549651205.

Stakem, Patrick H. *Microprocessors in Space*, 2011, PRRB Publishing, ISBN-1520216343.

Stakem, Patrick H. Computer *Virtualization and the Cloud*, 2013, PRRB Publishing, ISBN-152021636X.

Stakem, Patrick H. *What's the Worst That Could Happen? Bad Assumptions, Ignorance, Failures and Screw-ups in Engineering Projects, 2014,* PRRB Publishing, ISBN-1520207166.

Stakem, Patrick H. *Computer Architecture & Programming of the Intel x86 Family, 2013,* PRRB Publishing, ISBN-1520263724.

Stakem, Patrick H. *The Hardware and Software Architecture of the Transputer*, 2011, PRRB Publishing, ISBN-152020681X.

Stakem, Patrick H. *Mainframes, Computing on Big Iron*, 2015, PRRB Publishing, ISBN- 1520216459.

Stakem, Patrick H. *Spacecraft Control Centers*, 2015, PRRB Publishing, ISBN-1520200617.

Stakem, Patrick H. *Embedded in Space,* 2015, PRRB Publishing, ISBN-1520215916.

Stakem, Patrick H. *A Practitioner's Guide to RISC Microprocessor Architecture*, Wiley-Interscience, 1996, ISBN-0471130184.

Stakem, Patrick H. *Cubesat Engineering*, PRRB Publishing, 2017, ISBN-1520754019.

Stakem, Patrick H. *Cubesat Operations*, PRRB Publishing, 2017, ISBN-152076717X.

Stakem, Patrick H. *Interplanetary Cubesats*, PRRB Publishing, 2017, ISBN-1520766173 .

*Stakem, Patrick H. Cubesat Constellations, Clusters, and Swarms, Stakem,* PRRB Publishing, 2017, ISBN-1520767544.

Stakem, Patrick H. *Graphics Processing Units, an overview*, 2017, PRRB Publishing, ISBN-1520879695.

Stakem, Patrick H. *Intel Embedded and the Arduino-101, 2017,* PRRB Publishing, ISBN-1520879296.

Stakem, Patrick H. *Orbital Debris, the problem and the mitigation*, 2018, PRRB Publishing, ISBN-*1980466483*.

Stakem, Patrick H. *Manufacturing in Space*, 2018, PRRB Publishing, ISBN-1977076041.

Stakem, Patrick H. *NASA's Ships and Planes*, 2018, PRRB Publishing, ISBN-1977076823.

Stakem, Patrick H. *Space Tourism*, 2018, PRRB Publishing, ISBN-1977073506.

Stakem, Patrick H. *STEM – Data Storage and Communications*, 2018, PRRB Publishing, ISBN-1977073115.

Stakem, Patrick H. *In-Space Robotic Repair and Servicing*, 2018, PRRB Publishing, ISBN-1980478236.

Stakem, Patrick H. *Introducing Weather in the pre-K to 12 Curricula, A Resource Guide for Educators*, 2017, PRRB Publishing, ISBN-1980638241.

Stakem, Patrick H. *Introducing Astronomy in the pre-K to 12 Curricula, A Resource Guide for Educators*, 2017, PRRB Publishing, ISBN-198104065X.
Also available in a Brazilian Portuguese edition, ISBN-1983106127.

Stakem, Patrick H. *Deep Space Gateways, the Moon and Beyond*, 2017, PRRB Publishing, ISBN-1973465701.

Stakem, Patrick H. *Exploration of the Gas Giants, Space Missions to Jupiter, Saturn, Uranus, and Neptune*, PRRB Publishing, 2018, ISBN-9781717814500.

Stakem, Patrick H. *Crewed Spacecraft*, 2017, PRRB Publishing, ISBN-1549992406.

Stakem, Patrick H. *Rocketplanes to Space*, 2017, PRRB Publishing, ISBN-1549992589.

Stakem, Patrick H. *Crewed Space Stations,* 2017, PRRB Publishing, ISBN-1549992228.

Stakem, Patrick H. *Enviro-bots for STEM: Using Robotics in the pre-K to 12 Curricula, A Resource Guide for Educators,* 2017, PRRB Publishing, ISBN-1549656619.

Stakem, Patrick H. *STEM-Sat, Using Cubesats in the pre-K to 12 Curricula, A Resource Guide for Educators*, 2017, ISBN-1549656376.

Stakem, Patrick H. *Lunar Orbital Platform-Gateway*, 2018, PRRB Publishing, ISBN-1980498628.

Stakem, Patrick H. *Embedded GPU's*, 2018, PRRB Publishing, ISBN- 1980476497.

Stakem, Patrick H. *Mobile Cloud Robotics*, 2018, PRRB Publishing, ISBN- 1980488088.

Stakem, Patrick H. *Extreme Environment Embedded Systems,* 2017, PRRB Publishing, ISBN-1520215967.

Stakem, Patrick H. *What's the Worst, Volume-2*, 2018, ISBN-1981005579.

Stakem, Patrick H., *Spaceports*, 2018, ISBN-1981022287.

Stakem, Patrick H., *Space Launch Vehicles*, 2018, ISBN-1983071773.

Stakem, Patrick H. *Mars*, 2018, ISBN-1983116902.

Stakem, Patrick H. *X-86, $40^{th}$ Anniversary ed*, 2018, ISBN-1983189405.

Stakem, Patrick H. *Lunar Orbital Platform-Gateway*, 2018, PRRB Publishing, ISBN-1980498628.

Stakem, Patrick H. *Space Weather*, 2018, ISBN-1723904023.

Stakem, Patrick H. *STEM-Engineering Process*, 2017, ISBN-1983196517.

Stakem, Patrick H. *Space Telescopes,* 2018, PRRB Publishing, ISBN-1728728568.

Stakem, Patrick H. *Exoplanets*, 2018, PRRB Publishing, ISBN-9781731385055.

Stakem, Patrick H. *Planetary Defense*, 2018, PRRB Publishing, ISBN-9781731001207.

Patrick H. Stakem *Exploration of the Asteroid Belt*, 2018, PRRB Publishing, ISBN-1731049846.

Patrick H. Stakem *Terraforming*, 2018, PRRB Publishing, ISBN-1790308100.

Patrick H. Stakem, *Martian Railroad,* 2019, PRRB Publishing, ISBN-1794488243.

Patrick H. Stakem, *Exoplanets,* 2019, PRRB Publishing, ISBN-1731385056.

Patrick H. Stakem, *Exploiting the Moon,* 2019, PRRB Publishing, ISBN-1091057850.

Patrick H. Stakem, *RISC-V, an Open Source Solution for Space Flight Computers,* 2019, PRRB Publishing, ISBN-1796434388.

Patrick H. Stakem, *Arm in Space,* 2019, PRRB Publishing, ISBN-9781099789137.

Patrick H. Stakem, *Extraterrestrial Life,* 2019, PRRB Publishing, ISBN-978-1072072188.

*Stakem, Patrick H. Submarine Launched Ballistic Missiles,* 2019, ISBN-978-1088954904.

Patrick H. Stakem, *Space Command,* 2019, PRRB Publishing, ISBN-978-1693005398.

*Powerships, Powerbarges, Floating Wind Farms: electricity when and where you need it,* 2021, PRRB Publishing, ISBN-979-8716199477.

*Hospital Ships, Trains, and Aircraft,* 2020, PRRB Publishing, ISBN-979-8642944349.

2020/2021 Releases

*CubeRovers, a Synergy of Technologys,* 2020, ISBN-979-8651773138

*Exploration of Lunar & Martian Lava Tubes by Cube-X*, ISBN-979-8621435325.

*Robotic Exploration of the Icy moons of the Gas Giants*, ISBN- 979-8621431006.

*History & Future of Cubesats*, ISBN-978-1986536356.

*Robotic Exploration of the Icy Moons of the Ice Giants, by Swarms of Cubesats,* ISBN-979-8621431006.

*Swarm Robotics,* ISBN-979-8534505948.

*Introduction to Electric Power Systems*, ISBN-979-8519208727.

Patrick H. Stakem, *The Search for Extraterrestial Life,* 2019, PRRB Publishing, ISBN-1072072181.

*Centros de Control: Operaciones en Satélites del Estándar CubeSat* (Spanish Edition), 2021, ISBN-979-8510113068.

*Exploration of Venus*, 2022, ISBN-979-8484416110.

*Artemis Program*, 2021, ASIN-B09M5X6NVS; print version in work.

*James Webb Space Telescope*, 2021, ASIN-B09M5GR56Y, print version in work

www.ingramcontent.com/pod-product-compliance
Lightning Source LLC
Chambersburg PA
CBHW020914180526
45163CB00007B/2734